Volume 10 · 1990 · Number 1

Acta Biotechnologica

Journal of Biotechnology in Industry, Agriculture, Health Care, and Environmental Protection

Akademie-Verlag
Berlin

ISSN 0138-4988
Acta Biotechnol., Berlin 10 (1990) 1, 1—104

Instructions to Authors

1. Only original papers that have not been published previously will be accepted.
   Manuscripts may be submitted in English (in duplicate). The name of the institute (with the full address) from which the manuscript originates should be stated below the name(s) of the author(s). The authors are responsible for the content of their contributions.

2. Original papers should not exceed 20 typewritten pages (double spaced), including references, tables and figures; short original communications may contain a maximum of six typewritten pages.

3. Each paper should be preceded by a summary.

4. Latin names of species as well as passages to be printed in italics for greater emphasis should be marked by a waving line. Please use only units and symbols of the SI-system.

5. Tables may be used to shorten the text or to make it more comprehensible. They should be numbered consecutively throughout the text and be supplied with a brief heading. They should not appear in the text, but should be written on separate sheets.

6. The numbers and sizes of illustrations should be limited to a minimum, they should be numbered consecutively and be quoted on separate sheets.
   Line drawings, including graphs and diagrams, should be drawn in black ink. Half-tone illustrations should be presented as white glossy prints. Figure legends are to be typed in sequence on a separate sheet.
   The back of each sheet should bear the name(s) of the author(s).

7. References listed at the end of the contribution should contain only works quoted in the text. They should be numbered in the order in which they are first mentioned in the text.
   Please give surnames and initials of all authors, the name of the journal abbreviated according to "Chemical Abstracts — List of Periodicals", volume number, year of publication, issue number or month, first page number. Books are to be cited with full title, edition, volume number, page number, place of publication, publisher and year of publication.

8. Notes to the text may be presented as footnotes on the same page.

9. 50 offprints are free of charge.
   Additional ones may be ordered on payment.

10. The author will receive two galley proofs for correction. They are to be returned to the managing editor (Dr. Dimter, Permoserstr. 15, DDR-7050 Leipzig) as soon as possible.

Acta Biotechnologica

Journal of Biotechnology in Industry, Agriculture, Health Care, and Environmental Protection

Edited at the
Institute of Biotechnology of the Academy of Sciences of the G.D.R., Leipzig
by
M. Ringpfeil, Berlin and D. Pöhland, Leipzig

Volume 10

1990

Number 1

AKADEMIE-VERLAG BERLIN

"Acta Biotechnologica" publishes original papers, short communications, reports and reviews from biotechnology in industry, agriculture, health care and environmental protection.
The journal is to promote the establishment of biotechnology as a new and integrated scientific field.
The technological character of the journal is guaranteed by the fact that papers on microbiology, biochemistry, chemistry and physics must clearly have technological relevance.

Terms of subscription for the journal "Acta Biotechnologica"

Orders can be sent

- **in the GDR:** to Postzeitungsvertrieb or to the Akademie-Verlag Berlin, Leipziger Str. 3—4, PF 1233, DDR-1086 Berlin;
- **in the other socialist countries:** to a bookshop for foreign languages literature or to the competent news-distributing agency;
- **in the FRG and Berlin (West):** to a bookshop or to the wholesale distributing agency Kunst und Wissen, Erich Bieber oHG, Postfach 102844, D-7000 Stuttgart 10;
- **in the other Western European countries:** to Kunst und Wissen, Erich Bieber GmbH, General Wille-Str. 4, CH-8002 Zürich;
- **in other countries:** to the international book- and journal-selling trade, to Buchexport, Volkseigener Außenhandelsbetrieb der DDR, PF 160, DDR-7010 Leipzig, or to the Akademie-Verlag Berlin, Leipziger Str. 3—4, PF 1233, DDR-1086 Berlin.

Acta Biotechnologica

Herausgeber: Prof. Dr. Manfred Ringpfeil,
Akademie der Wissenschaften der DDR
Robert-Rössle-Str. 10, DDR-1115 Berlin-Buch und

Prof. Dr. Dieter Pöhland,
Institut für Biotechnologie der AdW der DDR
Permoserstr. 15, DDR-7050 Leipzig.

Verlag: Akademie-Verlag Berlin, Leipziger Straße 3—4, PF 1233, DDR-1086 Berlin;
Fernruf: 2236295 und 2236229; Telex-Nr.: 114420; Bank: Staatsbank der DDR, Berlin, Konto-Nr.: 6836-26-20712.

Stellv. Cheflektor Zeitschriften: Armin Beck; Redakteur der Abt. Zeitschriften: Cornelia Wanka.

Redaktion: Dr. Lothar Dimter (Chefredakteur),
Martina Bechstedt, Käthe Geyler,
Permoserstr. 15, DDR-7050 Leipzig; Tel.: 2392255.

Veröffentlicht unter der Lizenznummer 1671 des Presse- und Informationsdienstes der Regierung der DDR.

Gesamtherstellung: VEB Druckhaus „Maxim Gorki", DDR-7400 Altenburg.

Erscheinungsweise: Die Zeitschrift „Acta Biotechnologica" erscheint jährlich in einem Band mit 6 Heften. Bezugspreis eines Bandes 210,— DM zuzüglich Versandspesen; Preis je Heft 35,— DM. Der gültige Jahresbezugspreis für die DDR ist der Postzeitungsliste zu entnehmen.

Bestellnummer dieses Heftes: 1094/10/1.

AN (EDV) 18520

03000

Acta Biotechnologica

Journal of microbial, biochemical and bioanalogous technology

Edited by the
Institute of Biotechnology of the Academy of Sciences of the G.D.R., Leipzig
and by the
Kombinat of Chemical Plant Construction Leipzig—Grimma
by
M. Ringpfeil, Berlin and G. Vetterlein, Leipzig

Volume 9 · 1989

Annual index

Managing Editor: L. Dimter, Leipzig

ISSN 0138-4988 Acta Biotechnol., Berlin 9 (1989) 1—6, S. 1—578

AKADEMIE-VERLAG · BERLIN

"Acta Biotechnologica" publishes original papers, short communications, reports and reviews from the whole field of biotechnology.
The journal is to promote the establishment of biotechnology as a new and integrated scientific field.
The field of biotechnology covers microbial technology, biochemical technology and the technology of synthesizing and applying bioanalogous reaction systems.
The technological character of the journal is guaranteed by the fact that papers on microbiology, biochemistry, chemistry and physics must clearly have technological relevance.

Terms of subscription for the journal "Acta Biotechnologica"

Orders can be sent

- **in the GDR:** to Postzeitungsvertrieb or to the Akademie-Verlag Berlin, Leipziger Str. 3–4, PF 1233, DDR-1086 Berlin;
- **in the other socialist countries:** to a bookshop for foreign languages literature or to the competent news-distributing agency;
- **in the FRG and Berlin (West):** to a bookshop or to the wholesale distributing agency Kunst und Wissen, Erich Bieber oHG, Postfach 102844, D-7000 Stuttgart 10;
- **in the other Western European countries:** to Kunst und Wissen, Erich Bieber GmbH, General Wille-Str. 4, CH-8002 Zürich;
- **in other countries:** to the international book- and journal-selling trade, to Buchexport, Volkseigener Außenhandelsbetrieb der DDR, PF 160, DDR-7010 Leipzig, or to the Akademie-Verlag Berlin, Leipziger Str. 3–4, PF 1233, DDR-1086 Berlin.

Acta Biotechnologica

Herausgeber: Institut für Biotechnologie der AdW der DDR,
Permoserstr. 15, DDR-7050 Leipzig
(Prof. Dr. Manfred Ringpfeil) und

VEB Chemieanlagenbaukombinat Leipzig–Grimma,
Bahnhofstr. 3–5, DDR-7240 Grimma
(Dipl.-Ing. Günter Vetterlein).

Verlag: Akademie-Verlag Berlin, Leipziger Straße 3–4, PF 1233, DDR-1086 Berlin;
Fernruf: 2236201 und 2236229; Telex-Nr.: 114420; Bank: Staatsbank der DDR, Berlin,
Konto-Nr.: 6836-26-20712.

Redaktion: Dr. Lothar Dimter (Chefredakteur),
Martina Bechstedt, Käthe Geyler,

Permoserstr. 15, DDR-7050 Leipzig; Tel.: 2392255.
Veröffentlicht unter der Lizenznummer 1671 des Presseamtes beim Vorsitzenden des Ministerrates der Deutschen Demokratischen Republik.

Gesamtherstellung: VEB Druckhaus „Maxim Gorki", DDR-7400 Altenburg.

Erscheinungsweise: Die Zeitschrift „Acta Biotechnologica" erscheint jährlich in einem Band mit 6 Heften. Bezugspreis eines Bandes 198,— DM zuzüglich Versandspesen; Preis je Heft 33,— DM. Der gültige Jahresbezugspreis für die DDR ist der Postzeitungsliste zu entnehmen.

Bestellnummer dieses Bandes: 1094/9.

AN (EDV) 18520

Contents of Volume 9 (1989)

Number 1. 1989

Number 2. 1989

Number 3. 1989

Number 4. 1989

Number 5. 1989

Number 6. 1989

Acta Biotechnologica is indexed or abstracted in Current Contents/ET & AT;
Chemical Abstracts; Biological Abstracts; Biotechnology Abstracts; Excerpta Medica

Acta Biotechnol. 10 (1990) 1, 3–5 Akademie-Verlag Berlin

Biotechnology at the Treshold of the XXI Century – „Acta Biotechnologica“ is Entering its Second Decennium

RINGPFEIL, M.

Academy of Sciences of the G.D.R.
Röbert-Rössle-Str. 10, Berlin, 1115 G.D.R.

Scientific community is using the term Biotechnology since quite two decades. During this period a rapid development took place. The advance of biosciences into the molecular and cellular level of life processes, starting somewhat earlier, resulted in a principal understanding of structure and function of biomacromolecules, regulation and control of life processes, properties and processes of heredity, especially the identity and identical interpretation of the genetic code in all organisms. Simultaneously, the thermodynamics of life processes have been derived and the material and energetic base of life became conscious in a quantitative way.

These developments inaugurated an increasing interaction between biology and technology. The practical expression of this interaction is biotechnology. Its extension and progress is strongly dependent on its connections with biosciences, other natural sciences, and engineering sciences.

Inclusion of technical systems into biological in order to achieve new forms of life such as transgenic organisms and artificially altered proteins opened new horizons for utility and productivity of microorganisms, plants, and animals. The upcoming opportunities are not completely overlooked, borders of ethics and human responsibility seem to be touched. Microelectronics and micromechanics start to drive these inclusions beyond molecular levels in order to achieve stable forms of life support by micromachinery.

Inclusion of biological systems into technical, largely realized in traditional industries and services, receives considerable efficaciousness especially when artificially improved biological systems are enclosed. Also here ethical issues are involved. As a result of both developments creation of completely new processes as well as amplification and optimization of already available are in a heavy movement.

The molecular concept of biological research challenges a new development that may be called molecular bionics. It is aiming at realization of artificial systems for practical purposes simulating biological. Presumably it will grow as a competition for biotechnology offering more stability at comparable performance.

New techniques based on biological, chemical, and physical insights and applied to biological objects are accepted to be biotechnological. These are genetic engineering, hybridoma technique, protein engineering, biosensor techniques and bioelectronics. Systematic mastering of the great diversity of proteins and their connections to nucleic acids has been started by developing automated methods of analysis, synthesis, and data handling. More conventional techniques as cell culture techniques, immune techniques, and microbial, enzyme, and embryo engineering have been improved and amplified

considerably especially under the influence of genetic engineering and hybridoma technique, respectively.
Application areas of biotechnology have been proven to be health care, agriculture, industrial production, resources management, public services, and environmental protection. Basic demands of human and social life are met. The broad spectrum of biotechnological processes allows their application in different stages of economic and social development.
During the last years biotechnology went through its emancipation from being an industrial biotechnology only. The penetration into medicine and agriculture has opened broader perspectives for biotechnological development. As a result specific medical and agricultural branches of biotechnology have been accepted by medical and agricultural sciences. In ecological sciences it was recognized that the upcoming ecotechnology can develop only in concordance with a corresponding branch of biotechnology.
Facing the next decade it can be expected that biotechnology will develop as a science and as a field of practice in a strong connection with biosciences, other natural as well as engineering sciences. Progress will be influenced decisively by the ability of the biotechnologically working human potential to digest the last issues of natural and engineering sciences and to challenge the latter to create new. Involvement of genetic and protein engineering will increasingly determine the economic but also the moral value of the biotechnological applications and, therefore, influence the velocity and volume of the applications themselves.
Main applications of biotechnology will occur in medicine, pharmaceutical industry, and agriculture, and to a certain extent in food industry and environmental protection.
Efforts for developing a peripheric industry producing fine chemicals and devices will increase. Databanks will become a key for the development of genetic and protein engineering and their practical application.
Industrial production of commodity chemicals, certain food, feed, bulk pharmaceuticals and enzymes, as well as biological methods of making accessible raw materials and connecting waste disposal with recovery of secondary raw materials will not undergo great changes because of poor knowledge available at the molecular level of the biological systems serving as catalysts or being products of these industries. Otherwise, the already available processes will gain in importance as preferential means for broadening the base of materials, processes, and products of the developing countries. Scientific understanding of these processes has to be pushed forward as thight as possible to the modern developments of natural and engineering sciences.
Benefitting in developing countries from the achievements of biotechnology as well as from the modern developments of natural and engineering sciences for health care, agriculture, forestry, and environmental protection will grow as a principal need.
Biotechnological applications are affected with being involved in economic competition themselves. Competition will increase also from alternative non-biotechnological technologies stressing economy and technical realiability of biotechnological applications.
Biotechnological development and application will be influenced by the current programs of the leading countries as USA, USSR, and Japan as well as by the programs of the large economic blocs and by the efforts of effectively working international organisations like UNESCO, UNIDO, ICSU, EFB, and EMBO.
Ten years ago "Acta Biotechnologica" has been launched as an international scientific journal "dedicated to promote and support further development of biotechnology" [1]. Academy of Sciences and Ministery of the Chemical Industry of the G.D.R. supported this undertaking expecting a contribution to the development of both science and production. Eminent scientists and engineers from East and West did help promoting the exchange of results and ideas as members of the editorial board. Some of them passed

away. We remember W. Plötner, Leipzig, G. K. Skryabin, Moscow, and J. E. Zajic, El Paso.
The journal proved to be open for contributions from scientists working all over the world. Space was made available especially for our colleagues from developing countries. Publication could be amplified from 4 to 6 numbers a volume.
Now "Acta Biotechnologica" is stepping into the second decennium of its appearance. We take occasion to make a further step into the direction of modern biotechnology. We are willing to favourize contributions dealing with genetic engineering, hybridoma technique, protein engineering, biosensor techniques, and bioelectronics. We would like to accept papers emphasizing developments in the field of medical, agricultural, and environmental biotechnology. We are interested to present new developments in industrial biotechnology based on cell cultivation, use of extremophilic microorganisms, or new forms of enzymes. According to the general development of new technologies we would help forward systematic and scientific principles in biotechnology in face of an empirico-pragmatic approach applied yet frequently.
Better understanding is needed of what kind of science biotechnology is. "Biotechnology is not only Bio but also Technology" [2] and naturally "not only Technology but also Bio". Definition of biotechnology did not come to an end. We favour the opinion that the term biotechnology has a double meaning today. It stands for "an independent branch of applied sciences, based on natural and engineering sciences, directed to economic and social aims, developing, describing, and applying combinations of pragmatically selected biological and technical systems for productive, service, and life-supporting purposes" and for "a field of practice in which combinations of pragmatically selected biological and technical systems are used for productive, service, and life-supporting purposes". Can we overcome this double meaning by using the term "Biotechniques" for the field of practice?
In every case "Acta Biotechnologica" will contribute to spread knowledge about biotechnology as a science and help to consolidate biotechnology as a science. This seems to be the best way a biotechnological journal can go to promote the advance of biotechnical systems in practice for the final aim every scientific and practical effort should serve: the benefit of mankind.

Received June 14, 1989

References

[1] Scheler, W.: Acta Biotechnol. 0 (1980) 3.
[2] Ringpfeil, M.: Chem. Int. (IUPAC) 7 (1985), 19.

Book Review

M. E. Bushell

Progress in Industrial Microbiology

Vol. 25. Computers in fermentation technology

Amsterdam, Oxford: Elsevier, 1988. 198 pp., $ 73.75, Dfl. 140.00, ISBN 0-444-429794

In the last years we notice a rapid increase in the introduction and application of advanced computer hardware for fermentation control and of advanced software. In this volume two approaches to the complicated problems of the creation of software for fermentation control are described, one involving optimization statistics in fermentation process design and the other concentrating on the use of models for control processes.

After a brief introduction in the field of control strategies J. R. Court discusses in the first chapter "Computers in fermentation control: laboratory application" adaptive control algorithms for biological processes.

In the following part "Hardware and software strategies for bioreactor data acquisition and process control" S. C. Musgrave gives a review about computer systems and the implementation of suitable software for fermentation technology. Hardware and software configurations are considered in detail. Some aspects of problems of artificial intelligence and expert systems round up the chapter.

J. R. Leigh, J. Tampion and M. H. Ng inform about "Application of estimation techniques in fermentation technology". A suitable approach to the use of estimation techniques, especially of Kalman Filter method, is outlined in this chapter, drawing on own experience with Streptomyces and other fermentations.

In M. A. Winklers report "Optimization and time-profiling in fermentation processes" a number of advanced concepts are made accessible to fermentation scientists which are not necessarily interested in the underlying mathematical concepts. The author finishes with a section on time-profiled fermentation and the approach to optimizing genetically engineered yeast culture processes.

In the last chapter "Fermentation process development within a computer controlled pilot plant" T. Brix, S. W. Drew and B. C. Buckland give an impressive representation of a computer system, based on a Honeywell 4500 computer. Practically difficulties and solutions in the field of Streptomyces production processes are described.

This interesting edition with a lot of attractive details and many references describes several aspects of an area which is particularly exciting due to the number of unanswered questions, which still exist.

D. Bernhardt

Acta Biotechnol. 10 (1990) 1, 7–11 Akademie-Verlag Berlin

A Survey of Problems Relating to Legal Protection of Genetic Engineering Inventions

PEHU, T., GYLLENBERG, H. G.

Department of Microbiology, University of Helsinki, SF-00710 Helsinki, Finland

Summary

In the first patent applications in the field of genetic engineering the main problems were the terminology and the definition of various concepts. This was mainly due to the fact that the field of genetic engineering was new to the patent officials as well as to the patent attorneys. At the moment the terminology and the concepts have already become familiar and patent problems also in the microbiological sector are merely technical. However, new exiting problems have arisen because of the rapid development in the field of plant and animal biotechnology. A lot of work is still needed before the patent system fulfills its purpose as a promoter of technical, economical and social progress also in the field of biotechnology.

Introduction

Within the framework of scientific cooperation between the Academy of Finland and the Academy of Sciences of the USSR a joint project on the "Legal protection of genetic engineering inventions" has been carried out in the years 1982 to 1988. A number of documents have been produced in various stages of this project.

With this project a virgin field of research was approached. There are, of course, studies on the patent system and its implications, as well as on patents as indicators of trends and development areas in the R & D sector. In the case of this project, however, a distinctly new technology in an early stage of establishment has been analysed. There were no earlier models or methods available. The strategy of approach and the methods to be applied developed as new problems were faced.

The work was started with the investigation of the characterization of objects of genetic engineering. The purpose was to analyse the terms that are used in defining genetic material in patent applications and granted patents [1]. The work was later continued to investigate the development that had taken place in this respect [2]. These two studies represent the views of the patent applicants. Consequently, we became interested in the views of the patent authorities, which were elucidated by a case study based on the decisions given by the European Patent Office at the different steps of the patenting process [3]. Towards the end of the project special attention was paid to the questions concerning the patentability of the inventions in the field of plant molecular genetics, which is a rapidly expanding area of great importance in the field of genetic engineering

[4]. The problems related to the legal protection of genetic engineering and plant biotechnology inventions were more widely discussed in the surveys concerning the applicability of these inventions to the patent system [5, 6, 7, 8].
The main output of the above mentioned documents, produced by the Finnish party of the project, is summarized in this article.

Studies Carried out

Characterization of Genetic Engineering Inventions

In examination of patent applications in the field of genetic engineering one of the main problems, according to many patent experts, has been the terminology used in the applications for the characterization of the objects of inventions [9, 10, 11]. To analyse these problems two studies have been carried out [1, 2].
In the first study 235 patent documents in the field of genetic engineering, published before August 1982, were examined [1]. The examination showed that for the characterization of the "tools" of gene technology, i.e. plasmid, vector, recombinant DNA and DNA fragment, 12 primary characteristics and 7 secondary characteristics have been used. The list of 19 characteristics was also used in the examination of the research material from the years 1982—1986 [2].
The material of KNUTH et al. [2] comprised 439 patent applications published by the European Patent Office in the period 1982—1986. The main information collected from the application documents concerned the objects for which protection was claimed and the characteristics used in the claims to describe plasmid, vector, recombinant DNA and DNA fragment. Information about development of the number of applications, origin of the applications, products obtained by genetic engineering methods and deposition was also included.
The collected information about the objects of patent protection was compared to the data produced by SUETINA et al. [12] based on the European Patent applications published before 1982. The comparison showed that the objects of patent protection had not changed much. The results showed increase for claims dealing with microorganisms, vectors and DNA fragments. In general patent protection was more often claimed to products than processes for obtaining them. Most frequently patent protection was sought for microorganisms per se (72.2%). Obviously this reflects the increasingly positive attitude that patent authorities in many countries have shown towards patentability of "man-made" microorganisms.
According to this investigation the concepts used to define genetic material in patent claims were often ambiguous. For instance a vector may mean a plasmid to which an insert can be linked, or a plasmid to which an insert has already been linked. Also the contents of the concept recombinant DNA varied to some extent. However, when comparing applications from the period before 1982 [1] to applications published from 1982 to 1986, the terminology has perhaps developed in the direction of more distinctness.
There did not occur much difference in characterization of genetic material between the two studies. Most obvious differences could be observed in the descriptions of the nucleotide sequences and the regulatory regions. Complete nucleotide sequences of genes were presented somewhat more often than before. Plasmid and vector were more frequently characterized by promoter, less frequently by other parts of the regulatory region (operator, ribosome binding site etc.) than earlier. This is not surprising since a lot of work has lately been done for improvement of expression. As a general conclusion the description of genetic material had become more detailed and exhaustive.

Evaluation of Patentability in Genetic Engineering Inventions

In a case study based on 19 decided EP patent applications in the field of genetic engineering we extended our research from applications to the examination process of the applications [3]. The purpose of the study was to see, how the patentability of the inventions was evaluated by the European Patent Office and by opponents.
Out of the 19 cases, all of which EPO had reached a decision to the end of September 1986, in 14 cases a patent was granted and in five cases the applications were refused. Both for the granted patents and for the refused applications all documents relating to the different phases of the treatment process (official actions, oppositions etc.) were examined.
In Patent Office statements on accepted applications, most frequent topics have dealt with the description of the inventions, and their repeatability, novelty and inventive step.
In the majority of the applications the claimed patent protection was rather broad as compared to the corresponding descriptions. Especially applications, where inventions concerned protein production, belonged to this group. Usually patents have been granted without changes or with only minor changes as compared to the corresponding applications.
In the cases, where lack of novelty and inventive step have been reasons for a suggested restriction of the scope of protection, significant questions have been the evaluation of the equivalence of inventions and the assessment of what a person skilled in the art would consider obvious, which has turned out to be very problematic, because genetic engineering is a new field of technology.
The grounds for the refusal on the five applications varied significantly, as well as their contents and the claimed coverage. When comparing the decisions concerning the granted patents to the decisions concerning the refused applications the attitude of EPO seemed to have turned to a new, stricter direction.
Almost all the oppositions concerned patents with a broad protection and/or commercial significance (altogether six patents). The opponents consider the inventions questionable more often than the Patent Office and they are more concerned with the inventive step of the inventions. In four cases the patentees were willing to restrict the patent protection of already granted patents, to which oppositions have been raised.
Because of the scant material it was not possible to draw any far-reaching conclusions on the general principles behind the decisions of the Patent Office. However, patenting in the field of genetic engineering seemed still to be in a somewhat confused stage and that there was quite a difference between the views of the Patent Office and the users of the patent system.

Patent Documents Related to Plant Molecular Genetics

In the past several years a very large potential has developed in the field of genetic engineering for the transfer of the foreign genes to plants. This has created a need to improve the possibilities for the legal protection of the inventions in this field. TYNKKYNEN [4] carried out a survey to find out, what kind of inventions in the field of plant molecular genetics have been presented as patent applications, in which way the inventions have been verified in the descriptions, and what are the claimed objects of inventions.
The material of the investigation comprised 38 patent documents, filed in the period of 1979–1986. The patent documents could be classified into three groups according to the subject of the invention: Ti-plasmid vectors (*Agrobacterium*), plant virus vectors and protoplast technique.

The investigation showed that only in a few cases the patent documents described inventions having new remarkable changes in the characters of plants although that particularly is the aim of the new plant biotechnology. Only two documents showed inventions where new useful genes were expressed in regenerated plants (insecticide resistance in tobacco, herbicide resistance in potato). New genes transformed into plant cells comprised usually genetic markers.
A common feature for the verfication of the inventions was that transformed plant cells or plant tissues were not regenerated to whole plants. Plant experiments, regeneration of plants or transfer of foreign genes to a plant had been described in the specification part of the documents in considerable wideness as compared to the actual verification of the inventions in the example part.
According to the investigation the most work in plant biotechnology seems to have concentrated on improving the methods of gene transfer system. However, after the time of collecting the material, research in this field has made considerable progress and whole new plants with new beneficial characteristics have been developed.

Discussion

Principally any invention, which fulfills the patentability criteria: novelty, inventiveness and industrial applicability or utility, and which is reproducible, is patentable. However, applying old industrial property laws to new fields of technology like the new biotechnology, which deals with inventions concerning living matter, is difficult.
Describing a gene technological invention in a patent application in such a manner that a person skilled in the art can carry out the invention is often impossible. Although deposition of the tools of gene technology in an internationally recognized culture collection is largely recognized as a guarantee of reproducibility of the inventions, it is not always to the interest of the patentees, because of the variable deposit rules in different countries. For instance in United Kingdom and Federal Republic of Germany the deposited samples are publicly available from the preliminary publication onward [5, 7]. Nevertheless the applicants often deposit their samples only for safety, because it is easier than trying to convince the patent authorities of the reproducibility of the inventions. Obviously deposit practices should be internationally harmonized [5, 6].
In the first patent applications on genetic engineering the objects of the inventions were described in general terms and the claimed patent protection was rather broad [2]. The patent protection in the first granted patents could also be considerably broad compared to the description of the invention given in the specification [3]. In a new field of technology the patent protection can be broad, because prior art does not restrict the coverage of the patent. But if the invention does not function in that broadness as defined in the claims, the patent can be later revoked [5, 7].
On the other hand, if the patent protection covers only certain DNA constructions or transformed microorganisms, it may be easy for the competitors to use the invention without infringing the patent by preparing slightly different constructions and transforming them to other microorganism strains. To show the function of the invention in reasonable broadness has proved to be a bigger problem in biotechnological inventions than in the other fields of technology [5].
To fulfill the conditions of patentability, inventiveness and reproducibility may cause more problems than novelty and industrial applicability. The genetic engineering inventions are usually considered to be reproducible if the used biological material — donors, plasmids, recipients — are publicly available. The inventions are also usually industrially applicable because they concern often the production of commercially valuable protein

products by using microorganisms. But inventiveness may be questionable because genetic engineering methods are mostly well-known and not inventive [5].
In the field of biotechnology plant production is regarded as one of the most promising adaptations. Transformation of genetic information from one plant to another or the use of the bacteria as a donor are already well established methods in the art. In addition, it is relatively easy to produce great amounts of identical plant material for reproduction. The development of those methods is to a large extent on the level of basic research [4]. Due to the considerable investments required the development is concentrating in the research centers of the big multinational chemical companies. Due to this development there have been demands for improvement of the plant variety protection (UPOV convention) and for acquiring patents for plants.
It has tried to improve patent protection of plants by interpreting such typical law expressions as "a plant variety", "a microbiological method" and "an essentially biological method" in a new way and the patent law and its interpretations have been practicized all over the world with great flexibility. It is most evident that to maintain international competitiveness, patenting of plants shall become possible either by reinterpreting the law or by changing it [8].

Received January 27, 1989

References

[1] Knuth, S., Belyakova, L. G., Velkov, V. V., Dementyev, V. N., Lommi, H. J., Harper, A. L., Gyllenberg, H. G.: Acta Biotechnol. 4 (1984), 195.
[2] Knuth, S., Pehu, T., Gyllenberg, H. G.: World Patent Information 9 (1987), 229.
[3] Pehu, T., Knuth, S., Gyllenberg, H. G.: World Patent Information 10 (1988), 184.
[4] Tynkkynen, K.: Preprint in: Legal Protection of Genetic Engineering Inventions. Final report. Reports from Department of Microbiology, University of Helsinki. 36/1988.
[5] Knuth, S.: Geenitekniset keksinnöt ja patenttijärjestelmä. Biotekniikan ja käymisteollisuuden tutkimussäätiö. Helsinki. 1988. 63 p.
[6] Knuth, S.: E.B.C. — Symposium on brewer's yeast, Vuoranta (Helsinki), Finland. 1987. 209.
[7] Knuth, S., Gyllenberg, H. G.: Microbial Technology in the Developing World. Ed. E. J. DaSilva, Y. R. Dommergues, E. J. Nyns and C. Ratledge. Oxford University Press. 1987. 418.
[8] Vanhala, P.: Preprint in: Legal Protection of Genetic Engineering Inventions. Final report. Reports from Department of Microbiology, University of Helsinki, 36. 1988.
[9] Crespi, R. S.: Patenting in biological sciences. John Wiley & Sons. Chichester. New York. 1982. 211 p.
[10] Crawley, P. E.: Genetic Engineering (Part I). Patent protection in biotechnology, 28th June 1985. European Study Conferences Ltd. C1F-23F.
[11] Biotechnology and Patent Protection. An international review. Ed. F. K. Beier, R. S. Crespi and J. Straus. Organization for Economic Co-operation and Development. Paris. 1985. 133 p.
[12] Suetina, R. L., Belyakova, L. G., Velkov, V. V.: Biotechnology and recombinant DNA. Scientific Center of Biological Research of the Academy of the Sciences of the USSR in Pushchino. 1982. 237 p.

Book Review

M. F. TUITE, M. PICCARD, M. BOLOTIN-FUKUHARA (Herausgeber)

Genetics of Translation

New Approaches. NATO ASI Series H: Cell Biology, Vol. 14.

Berlin, Heidelberg, New York: Springer-Verlag, 1988. 524 S., 145 Abb., DM 248,—
ISBN 3.540.18550-X

Unter diesem anspruchsvollen Titel findet sich die Sammlung der auf einem Symposium in Aussois vom 21.—26. Mai 1987 gehaltenen 40 Vorträge. Ein ausgewählter Teilnehmerkreis, zur einen Hälfte aus den USA, zur anderen aus Europa, war der Einladung zu einem NATO-Workshop in die französischen Alpen gefolgt und suchte neue Ansätze zur Genetik der Translation auf der Basis der vorgestellten eigenen Untersuchungsergebnisse zu finden. Nachdem in den letzten 20 Jahren die strukturellen Grundlagen der am Translationsprozeß beteiligten einzelnen RNA- und Proteinmoleküle ihrem Wesen nach aufgeklärt worden sind, war es Hauptanliegen und erklärtes Ziel der Veranstalter, die Möglichkeiten des Einsatzes genetischer Techniken zur Untersuchung der Regulation des Translationsprozesses zu diskutieren. Dieses Ziel wurde — nach Meinung der Organisatoren — in kreativer Atmosphäre erreicht; dem Leser der im Kongreßband abgedruckten Einzelbeiträge bleibt dieser Eindruck allerdings verborgen, da die sicher interessanten Diskussionen hierzu nicht angefügt wurden.

Die Vorträge sind in die 7 Sachgebiete ribosomale RNA, ribosomale Proteine, Transfer-RNA, Translationsfaktoren, Mechanismus der Translation, Genauigkeit (accuracy) der Translation und Translationskontrolle aufgegliedert. Bedenkt man, daß auf jedem dieser Sachgebiete bedeutende Unterschiede zwischen Pro- und Eukaryonten — etwa 60% der Beiträge befaßten sich mit Bakterien, 30% mit Hefe und der Rest mit anderen Eukaryonten — bestehen, so liegt auf der Hand, daß gegenwärtig die „Genetik der Translation" noch ein sehr heterogenes, von vielen Einzelbefunden geprägtes Wissenschaftsgebiet darstellt. Die Beiträge des Buches enthalten interessante Einzelergebnisse, vor allem zu funktionellen Aspekten der Translation; auf strukturelle und mechanische Betrachtungen zum Ribosom wurde — schon durch die Wahl des Teilnehmerkreises bedingt — bewußt verzichtet. Es fällt schwer, einzelne Artikel hervorzuheben, ohne gleichzeitig andere zurückzusetzen.

Zur Kennzeichnung des breiten Spektrums der behandelten Themen sei erwähnt, daß in 12 Beiträgen über Ergebnisse berichtet wird, die nach ortsgerichteter Mutagenese erhalten wurden. Dabei handelt es sich um Untersuchungen an ganz unterschiedlichen Zellsystemen, so bei *Escherichia coli* um die Synthese von rRNA, Tryptophansynthetase, Elongationsfaktor EF-Tu, sowie von Proteinen des Bakteriophagen MS2, bei Hefe um die Phosphorylierung des ribosomalen Proteins S 10 oder auch der Rolle der Initiationskodons verschiedener Messenger-RNAs, bei Säugerzellen um die Rolle des Initiationsfaktors eIF-2alpha oder auch in einem aus *Drosophila melanogaster* isolierten System um die $tRNA^{Tyr}$-Genexpression.

Zur weiteren Information über die einzelnen behandelten Themen werden umfangreiche Literaturhinweise gegeben, des weiteren enthält das Buch ein Stichwortverzeichnis. Es sei auch erwähnt, daß die Wiedergabe der Offsetarbeiten einschließlich der darin enthaltenen Abbildungen von guter Qualität ist.

Für Naturwissenschaftler, die in der molekulargenetischen Grundlagenforschung tätig oder an ihr interessiert sind, kann das Buch wertvolle Anregungen vermitteln.

J. STAHL

Acta Biotechnol. 10 (1990) 1, 13—22 Akademie-Verlag Berlin

Use of Antibodies to the Lipopolysaccharide Core Region in the Treatment of Gram-negative Sepsis and Shock

Appelmelk, B. J.[1]; Verweij-van Vught, A. M. J. J[1]., Brade, H.[2], Cohen, J.[3], Maaskant, J. J.[1], Thijs, L. G.[4], Mac Laren, D. M.[1]

[1] Department of Medical Microbiology, Vrije Universiteit,
Van der Boechorststraat 7,1007 MC Amsterdam, The Netherlands
[2] Institute of exp. Biology and Medicin
Parkallee 1—40, 2061 Borstel, F.R.G.
[3] Royal Postgraduate Medical School,
Hammersmith Hospital, London UK.
[4] Department of Acute Internal Medicine, Vrije Universiteit,
De Borlelaan 1117, 1007 MB Amsterdam, The Netherlands

Summary

In this paper we review evidence in favor of a protective role for antibodies directed at the lipopolysaccharide (LPS) core region of Gram-negative bacilli. It will be shown that, given the right animal model, polyclonal antisera raised against O-antigen lacking, rough mutant strains, can be shown to protect animals against lethal sepsis due to a variety a bacterial species. Evidence for a protective role obtained from clinical studies will also be discussed. It will be stressed how difficult it is to prove that the antibodies directed against the LPS-core, and not other proteins present in the polyclonal anti-rough strain antisera, were responsible for the observed protective effects. The solution to some of these problems is the production of monoclonal antibodies (Mabs). We will discuss the findings that purified anti-core Mabs may give false-positive outcomes of protection studies, due to the induction of tolerance by small amounts of contaminating endotoxin, present in the antibody preparation. The need to administer the Mab therapeutically instead of prophylactically to the test animals, will be stressed. It will be shown how we succeeded in determining the epitope-specificities of several anti-core antibodies. It will be shown also that use of Elisa alone for characterization of anti-core Mabs may lead to false conclusions concerning epitope specificities. We will prove that synthetic KDO-(=ketodeoxyoctanate) and lipid A-containing antigens are indispensable tools for thorough characterization of Mabs. Using this methodology we were able to discriminate between specific antigen-antibody interactions, and the ability of some Mabs to bind "non-specifically" to hydrophobic surfaces. The relationship between epitope specificity and protective effects of anti-core Mabs will be described. Finally, we will demonstrate that a KDO-specific Mabs is capable of protecting mice against lethal sepsis; in other words, it will be shown conclusively that Mabs to defined epitopes in LPS core region are cross-protective.

Introduction

Gram-negative sepsis and shock are very serious clinical problems: their incidence has been rising dramatically during the last 30 years from less than 1 case per 1000 hospital admissions in 1950, through some 8 in 1965, towards 13 in 1980. A recent estimate for the incidence of this disease state in the USA is 200,000 patients per year of whom some 75,000 die [1]. It has been well established that the actual mortality is strongly dependent on the intactness of host defence: in otherwise healthy patients, the mortality is about 15%; in severely immunocompromised patients this figure may rise to a staggering 85%. An example of the latter group are patients suffering from neoplasia, in particular when under treatment with cytostatic agents or following radiation therapy.

The incidence of infections has been shown to correlate strongly with the degree of neutropenia. The ever increasing number of hospitalized, immunocompromised patients, in combination with the greater sensitivity of Gram-positives for antibiotics, largely accounts for the increase in incidence of Gram-negative sepsis observed in the period 1950—1980. In addition, many invasive support or diagnostic procedures are now part of modern hospital practice; an example is the increasing use of intravascular catheterizations that inevitably will lead to more infectious problems.
It is now well established that bacterial endotoxins, also called lipopolysaccharides (LPS) [2, 3], are key substances in the aetiology of Gram-negative septic shock. One of the reasons why antibiotic treatment fails so frequently, is the inability of antibiotics to neutralize endotoxin. Notable exceptions with regard to neutralization of LPS are antibiotics of the polymyxin group but these are considered to be too toxic for general use. Thus, the more recently developed potent antibiotics lead to an efficient killing of the invading microorganisms, but the endotoxin stays unaffected and is still capable of producing potentially lethal effects. One of the possibilities of increasing host defence would be to prepare anti-LPS antisera, which are able to reduce mortality of Gram-negative sepsis. Indeed, such antisera, most often obtained by immunizing rabbits with whole killed cells or with purified endotoxin, were highly effective in lowering the mortality of experimental Gramnegative sepsis. However a serious limitation to their use was encountered: an antiserum raised against a given strain was almost exclusively active against that particular strain and not against others. This specificity can be explained as follows: LPS consists of 3 structural regions, i.e the lipid A part which is the toxic part, a so-called core region which is an oligosaccharide connected at one end covalently to lipid A, and at the other end to the third region, the O-antigen. The O-antigen is a polysaccharide consisting of repetitive units of oligosaccharide building blocks (see Fig. 1).

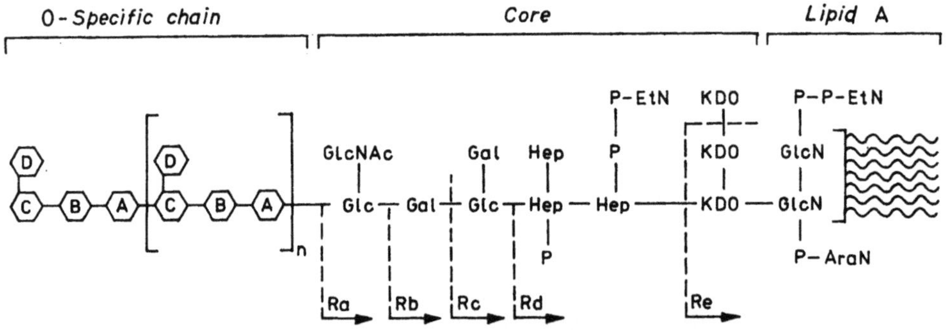

Fig. 1. General structure of a *Salmonella* lipopolysaccharide; A—D, sugar residues; Glc, D-glucose; Gal, D-galactose; GlcN, N-acetyl-D-glucosamine; Hep, L-glycero-D-manno-heptose; KDO, 2-keto-3-deoxy-manno-octonate; AraN, 4-amino-L-arabinose; P, phosphate; EtN, ethanolamine; ~~ hydroxy- and non-hydroxy fatty acids. Ra to Re, indicated are incomplete R form lipopolysaccharides. (Modified from Lüderitz et al., Pharmac. Ther. 15:383—402 (1982))

When whole killed cells or purified endotoxin are used as a vaccine, a strong immune response takes place, in particular to the O-antigen, and much less to lipid A or the core.
Hundreds of structurally different O-antigens are known among the three Gram-negative pathogens which predominate in sepsis and shock, i.e. *Escherichia coli*, *Klebsiella pneumoniae* and *Pseudomonas aeruginosa*. Consequently, the above mentioned anti-O antisera are active only against strains of the same O-serotype.

Thus, for adequate serotherapy, one would need a huge range of specific antisera. Within the clinical setting the use of type-specific antisera for treatment of septic patients is subject to a great practical difficulty: in general it takes several days before the invading microorganism is identified and serotyped, both being essential diagnostic procedures when type specific antisera are used for therapy. However, the often acute nature of Gram-negative sepsis and shock requires immediate therapeutic intervention.

A possible solution to these difficulties would be the development of a universal antiserum active against all or the most important Gram-negative pathogens. To produce such an antiserum two approaches are possible: a) preparation of the many different monospecific antisera for use in a serum pool, and b) preparation of an antiserum, raised against a structure which is common to the various Gram-negative species.

The first approach is considered too monumental a task to accomplish. A variant to this procedure is the screening of bloodbank donor units for anti-O antibodies; when suitable donor units are pooled, a product may be obtained containing high titers against a wide variety of O-antigens. The antiserum thus prepared was shown to be effective in several experimental models for endotoxemia [4] but data concerning clinical efficacy are unfortunately rather limited [5].

Much attention has focussed on the second approach i.e. the preparation of a cross-protective antiserum raised against a structure common to all Gram-negatives, and therefore capable of giving protection against the lethal effects of challenge with heterologous bacteria or LPS. The final choice of target for such a broadly reactive serum was the LPS core-lipid A region, because chemical and immunological investigations had clearly shown that these two components of LPS, in contrast to the O-antigen, contained structural elements common to a variety of Gram-negative species [6]. Thus, at least theoretically, antisera to core or lipid A could be expected at least to react with many strains. Of course it was hoped that such an antiserum would also be cross-protective against the most important pathogens. However, as mentioned above, the LPS core region is only weakly immunogenic, the O-antigen being immunodominant. The problem of preparing an antiserum with a high titer of antibodies to the core was solved by using so-called rough mutant strains as a vaccine. These mutants differ from their parent strain in not having an O-antigen. When the O-antigen is lacking, the core becomes strongly immunogenic, and thus when rough strains are used as a vaccine, the antiserum obtained contains large amounts of anti-core antibodies.

Many different rough mutants have been isolated, which differ from each other in the actual length of the core structure: i.e. some mutants contain the complete core region and this group of mutants is designated as the Ra-chemotype; those mutants with the shortest core are designated as Re-chemotypes. A Rc mutant has a LPS with an intermediate length, etc. Fig. 1 illustrates this principle.

Most of the work has been done with an *E. coli* rough mutant of the Rc-chemotype, called J5, and a rough strain of *Salmonella minnesota* of the Re-chemotype, called R595. Several research groups started with the preparation of such polyclonal antisera (most often raised in rabbits) to rough mutants, and with their testing for protection against lethal doses of live bacteria or purified endotoxin. The results of protection studies in laboratory animals were highly confusing: while some groups consistently obtained positive results (i.e. protection as a result of immunization with anti-rough strain antisera), many others were not successful. The relevant literature has been reviewed recently [7]. Possible explanations for these discrepant outcome could be: use of different vaccines (J5 or R595), heat-killed vs. acetone-killed vaccines; use of different vaccination schedules and routes; the inability to compare the antibody titers obtained by the various investigators, due to use of different assay methods; use of different routes for passive immunization and challenge; use of different challenge strains and LPS preparations; use of different mouse strains, and finally, the use of different animal models. We think that in particular the latter parameter is of great importance and that the discrepant outcomes may be in part explained by the choice of model. In short, two types of model have been used: 1) non-compromised mice in which very high challenge doses of LPS or bacteria are required for lethality: 0.2 mg LPS or $> 10^7$ CFU of bacteria, and 2) compromised mice where lethal doses may be very low: 1 ng LPS or < 1000 CFU. To obtain lethality at such low doses, mice must be compromised by treatment with enhancing agents such as mucin, hemoglobin, galactosamine and actinomycin D. Review of the literature [7] made clear that successful protection with anti-core antisera was much more often obtained in the compromised mouse model than in the "normal" mouse model. These two models were also directly compared (i.e. with the same batch of anti-core antiserum) both by us [8] and others [9]. In both studies an anti-core antiserum provided no protection in non-compromised mice, while protection was found in mice treated with mucin and/or hemoglobin. Compromised models are more sensitive, probably because the ratio of protecting antibody to challenge dose is higher than in non-compromised mice.

An additional problem was that in no single successful study sufficient evidence was given to prove that indeed the anti-core antibodies were responsible for the protective effects observed. The protection observed might have been due (besides to anti-core antibodies) to each one of the following factors: 1) Endotoxin contamination of the antiserum; this may cause, in very low doses, a non-specific increase in resistance. 2) Antibody response to surface structures (e.g. outer membrane proteins), other than rough LPS in cases where the vaccine consisted of whole cells instead of LPS. 3) The presence in the anti-rough strain antiserum of anti-O antibodies due to LPS-induced polyclonal B-cell activation [10]. 4) The presence of nonantibody proteins in antisera, like LPS binding proteins [11], acute phase proteins and/or mediators. All these factors could have contributed to the protection observed; it is thus evident that a rigorous proof is required that the anti-core antibodies present in a polyclonal anti-rough strain antiserum were indeed responsible for the observed protection. In a very limited number of studies hard evidence for the role of anti-core antibodies was sought by absorbing the antiserum with cellular vaccine, or with rough LPS [9, 12, 13], but as far as we know, in no single case was it demonstrated that the absorption procedure led to loss only of the anti-core antibodies, and not of other proteins or antibodies.

In summary, although there is quite some evidence in favor of a protective role for anti-core antibodies as present in polyclonal antisera, hard proof is currently not available. For this reason, many investigators have turned to the production of monoclonal instead of polyclonal anti-core antibodies (see below).

A second line of evidence in favor of a protective role for anti-core antibodies was obtained from clinical studies: in three studies, anti-core antibody titers were measured in sera from patients in the acute phase of sepsis/shock [14—17]. The conclusion was that those patients with a high anti-J5 or anti-R595 LPS titer were more likely to survive than patients with low titers in their serum. These data are strongly suggestive for a decreased mortality thanks to protection provided by anti-core antibodies. However, by nature, such studies demonstrate a correlation, which is not by necessity equal to a cause-and-effect relation.

A third line of evidence for protection mediated by anticore antibodies are clinical trials. In the first trial reported [18], healthy volunteers were immunized with J5, and the non-pooled units were transfused to patients already in sepsis/shock. The trial was double blind, the controls being donor units obtained before immunization. The result was that those patients who received immune serum/plasma had a much higher survival than those which received control plasma/serum. A second trial [19] in which anti-core antiserum was given before the onset of sepsis/shock did not show any favorable effects on mortality. However, this second trial is very difficult to evaluate statistically due to the small number of patients involved, and the low number of septic periods/deaths observed. In a third trial, non-pooled donor units were given to ICU patients at risk for infection [20]. A statistically significant protective effect was observed due to the transfusion with anti-J5 antiserum. A critical remark may be that for some statistical calculations very small numbers of patients were compared, with in one case a p-value of 0.049. In a fourth trial [21], pooled anti-J5 IgG was compared with a more traditional i.v. applicable immunoglobulin preparation. No benefit of the anti-J5 preparation could be demonstrated, For this trial the following remarks are of relevance: the anti-J5 IgG contained only a marginally higher anti-core antibody titer than the control Ig. Furthermore no data are available concerning its protective abilities in experimental sepsis. Lastly, the failure to protect might have been due to the fact that the anti-J5 preparation did not contain IgM. More and more data is currently accumulating that IgM anti-J5 (R595) and not IgG is protective [22].

It is clear that the outcomes of clinical trials, like those of animal studies are discrepant. In addition, the following remarks may be made: on the assumption that anti-core antibodies are protective, the actual anti-core antibody titer already present in the serum of patients before immune therapy is likely to influence the outcome (i.e. survival or death). Theoretically, a bad match of treatment and control group with regard to pre-existing anti-core titers could have influenced the outcome of the trials, be it in favor of protection or the opposite. We have already discussed the huge problems in demonstrating unequivocally that protection against experimental sepsis was indeed mediated by the anti-core antibodies themselves. Clearly, obtaining such hard proof from data of clinical trials where very complex and heterogenous groups of patients are compared, is even less likely. Some of the problems associated with abovementioned clinical trials may be solved as follows: one could screen bloodbank donor units for high titers of naturally occurring anti-core antibodies. Pooling of those units with the highest titers would lead to a single product (in stead of the often followed procedure of transfusing non-pooled units containing widely varying titers), with a higher titer than obtained by active immunization with J5 or R595. A product obtained in this way containing high IgM-anti J5 titers has already been marketed under the name "Pentaglobin" [23]. Its efficacy in experimental sepsis has been documented but no large, double-blind clinical trial has as yet been reported on.

Given the problems of proving protection by polyclonal anticore antibodies, it was no more than logical that many investigators have now turned toward the preparation of monoclonal antibodies (Mabs). When we ourselves started to prepare Mabs to the LPS core region we defined two major questions to be answered:

1. Are anti-core Mabs capable of neutralizing homologous LPS (e.g. the combination of anti-R595 Mabs and R595 LPS), and can we understand why. For instance, theoretically one would expect that antibodies with binding sites very near to the toxic part of LPS (i.e. lipid A) would be more likely to protect than antibodies binding further away. This question thus refers to the actual mechanism of neutralization without reference to possible cross-reactive/cross-protective abilities.

2. Are anti-core antibodies capable of cross-protection, and are we capable of defining the properties which make a certain antibody cross-protective, e.g. do cross-protective antibodies belong only to a certain immunoglobulin class, and/or are they very limited in their range of epitope-specificities?

In order to answer both questions, Mabs against LPS of R595 were prepared [24]. R595 has two advantages over J5: while the covalent structure of J5 has only been partly elucidated [25], that of R595 (and of some other rough mutants of *S. minnesota*) is known now mainly due to the work of BRADE and co-workers [26]. In addition, many fully defined, chemically synthesized lipid A and KDO containing structures are available now [27—29], which are indispensable in the elucidation of epitope-specificities (see below). It must be clear that a detailed knowledge on the epitope-specificities of the Mabs prepared is a sine qua non for answering the two major questions posed above.

Balb/c mice were immunized with a variety of R595 vaccines; several immunization protocols were followed. By routine procedures 12 Mabs, highly reactive with R595 LPS, were selected from a total of 10,000 R595 LPS-positive culture wells. To answer the first question, ascites was mixed with increasing doses of R595 LPS and injected into actinomycin D-sensitized mice. 6/12 Mabs gave significant protection, and up to a 90-fold increase in LD50 was observed [30]. No correlation between ability to protect and any of the following parameters was observed: isotype, antibody titer, antibody affinity. Thus, protective as well as non-protective clones belonged to various subclasses, and some low-titered ascites preparation were protective, while some very high titered ones were not. These negative findings led us to investigate whether or not there was a correlation between the binding site of the Mabs and their ability to neutralize R595 LPS. To this end, the reactivity of the Mabs with a variety of rough LPS was determined. Very defined reactivity patterns were observed: e.g. clone 17 reacted only with LPS of the Re-chemotype, while clone 25 reacted as well with Re and Rb2 LPS; a third clone (clone 20) exhibited the broadest cross-reactivity: it reacted with a broader panel of *S. minnesota* rough LPS but also with LPS of *Chlamydia* [31].

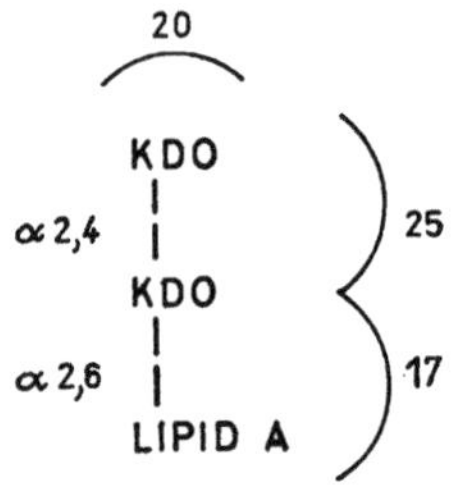

Fig. 2. Epitope specificities of three monoclonal antibodies (clones 17, 20 and 25) to LPS of *Salmonella minnesota* strain R595

Based on knowledge of the covalent structures of the LPS tested, a hypothesis concerning the various binding sites could be put forward. Definite proof of the hypothesis could be obtained by use of synthetic core/lipid A structures [32]. The binding sites of clones 17, 20 and 25 are shown in Fig. 2. Clone 20 is specific for α 2.4-linked KDO monosaccharide, clone 25 for α 2.4-KDO-KDO disaccharide and clone 17 recognizes a more complex structure: for binding of clone 17 KDO and part of lipid A is required [32, 33]. Of importance is that clones 17 and 20 were found to increase the LD50 in actinomycin D treated mice, following challenge with R595 LPS [30], while clone 25 had no effect. From the data shown in Fig. 2 at least one definite conclusion can be drawn: R595 LPS neutralizing Mabs (e.g. 17 and 20) do not have the common property of binding to the same epitope. Fig. 2 shows that the binding site for the non-protective clone 25 is located in between those of 17 and 20. However, we must bear in mind that in Fig. 2 only a two-dimensional representation of R595 LPS is given. Theoretically, the three-dimensional structure of R595 LPS could be such that the protective clones (17 and 20) do have a common property, i.e. a spatial closeness of their binding sites, with that of clone 25 located further away. This possibility was tested by using the LPS core-binding antibiotic polymyxin B as a probe for the three dimensional structure of R595 LPS. However, also the data thus obtained were not in favor for a close spatial proximity of the binding sites of clones 17 and 20 [34]. We must conclude that despite the use of the fully defined R595 LPS for challenge and of antibodies with known epitope specificities

for protection studies [32, 33], we are not able at present to understand why some antibodies are neutralizing while others have no effect. In contrast, on the cellular level we now have more insight how clone 17 could exert its protective effect: clone 17 is capable of inhibiting R595 LPS-induced lymphocyte activation *in vitro* [35]. It is possible that *in vivo* a similar process takes place and that clone 17 protects by diminishing LPS-induced secretion of toxic mediators like TNF or Interleukins.
To answer the question wheter anti-core Mabs are capable of cross-protection, the follow-wing experiments were done: at first, the 12 R595 LPS Mabs were investigated for their ability to bind with heterologous smooth LPS and bacteria (*Escherichia coli, Klebsiella pneumoniae, Pseudomonas aeruginosa*). Only one clone (no. 20) had the ability both to neutralize R595 LPS and to bind significantly to heterologous LPS/bacteria [36]. 10 µg clone 20 was able to protect mice against challenge with a smooth *E. coli* strain. When given prophylactically, it reduced the mortality from 72% to 0% [36]. Similar data were obtained when the antibody was given after challenge. This latter finding is of crucial importance: when antibodies are given prophylactically the possibility exists that small amounts of endotoxin, present in the Mab preparation, will cause false positive protective effects due to non-specific increase of resistance (tolerance). The answer to the second major question posed at the onset of our studies, i.e. are anti-core Mabs capable of cross-protection, was thus given.
We will now discuss in short some studies on other cross-protective Mabs. Three other cross-protective Mabs have been described in the literature:

1. Scientists at Stanford University and at San Diego have developed a human IgM Mab, prepared from spleen cells of an J5-immunized donor [37]. This Mab (called HA-1A) was further developed by CENTOCOR Inc. It was found to protect mice against challenge with the three relevant bacterial species when the antibody was given prophylactically [37]. It was also protective against the dermal SHWARTZMAN reaction in rabbits [37], as well as against lethal sepsis in neutropenic rabbits when given therapeutically [38]. Favorable effects were seen when given to rats, in a model were LPS was injected to elicit ARDS-like symptoms (ARSD = adult respiratory distress syndrome) [39]. This Mab is currently being tested for clinical efficacy.

2. A second human IgM Mab was prepared by LARRICK et al. at the CETUS Corporation [40]. It was prepared using peripheral lymphocytes of a non-immunized donor, with a high serum titer of anti-core antibody, as a source of immune B cells. This clone, coded D234, was protective against sepsis due to *E. coli*, when given therapeutically [41], and also against some of the deleterious effect of *E. coli*-induced meningitis in rabbits [42]. Clinical testing will soon begin.

3. A murine IgM anti-J5 Mab was prepared by L. S. YOUNG [43]. This Mab was protective when given prophylactically to mice challenged with *P. aeruginosa*. This Mab, coded XOMEN-E5, was further developed by the XOMA Corporation. Data concerning its pharmacokinetics have been published [44]. A clinical trial using this clone was completed very recently [45], but data concerning its clinical efficacy are not available yet.

It is very clear, both from our own studies and those of other scientists, that cross-protective Mabs do indeed exist. These findings put an end to the many disputes which have risen when polyclonal anti-core antibodies were used. How broadly cross-protective the various Mabs are is not yet clear: e.g., although HA-1A was shown to be protective against *E. coli, K. pneumoniae* and *P. aeruginosa*, when given prophylactically [37], LPS-induced tolerance could have mimicked the observed protection. It has been shown that even amounts of contaminating LPS as low as some nanograms can induce tolerance [46, 47].

What then are the general rules that determine whether or not a given Mab is cross-protective? It is striking that the four cross-protective Mabs discussed above are all of the IgM class. The evident question is: is that a coincidence or are IgG Mabs simply not able to cross-protect? If so, what is it in IgM that makes the crucial difference?

Possibly it is important that the multipoint attachment of pentameric IgM to the multivalent LPS antigen (LPS forms multivalent aggregates when in solution) allows a much more avid binding than in case of an IgG Mab.
Another crucial point is whether or not these four Mabs have identical or similar epitope specificities. Unfortunately this question cannot be answered yet: only clone 20 has been characterized in detail in this respect [32, 33] and as demonstrated above it is specific for KDO, and not for lipid A. The other three clones [37—45] have been shown in Elisa to react with lipid A as well as with various rough LPS. At first sight one would classify them as lipid A-core Mabs, i.e. the epitope would consist of parts of lipid A and the core. We have found that many of our anti-R595 LPS (e.g. clones 17, 20 and 25) Mabs reacted in Elisa with lipid A and rough LPS [36]. From our subsequent studies it has become clear that clones 17, 20 and 25 all were of a different epitope specificity. Thus, use of Elisa for binding studies may give the false impression that various Mabs have similar or identical epitope specificities. Therefore at present it is not clear whether or not the various cross-protective Mabs bind to the same epitope. For thorough characterization of binding sites, synthetic core antigens are indispensible. Another important feature came out when we used synthetic antigens: although clone 20 does bind in Elisa to lipid A, no such reactivity was seen using the passive hemolysis technique. In this system, LPS (or lipid A-) -coated SRBCs are allowed to react with antibody, following which complement is added. Lysis takes place when the bound antibody fixes complement. Clone 20 does lyse R595 LPS coated SRBCs, but not lipid A-coated ones. In combination with the outcomes of reactivity studies using synthetic KDO antigens we want to postulate that binding of a Mab in Elisa to lipid A does not necessarily equals its epitope specificity. Therefore, the binding in Elisa of some Mabs (e.g. clone 20) to lipid A reflects their "stickiness", and not their epitope-specificities. Therefore it is possible that also clones HA 1A, XOMEN-E5 and D234 are specific for KDO and not for lipid A.
Summarizing, for the first time it has now been demonstrated that an antibody specific for KDO (clone 20) does exhibit cross-protective behavior [6]. However, by what mechanism crossprotection takes place is not known yet. We expect that understanding of the mechanism will lead to preparation of Mabs even more effective than the ones currently known. We also expect that use of these Mabs within the clinical setting will soon belong to the accepted armamentarium of modern ICU's.

Received January 31, 1989

References

[1] Hamill, R. J., Maki, D. G.: — In: Handbook of endotoxin, Vol. 4, p. 55, Proctor RA, Ed. Clinical aspects of endotoxin shock. Amsterdam, Elsevier (1986).

[2] Rietschel, E. Th., Brade, L., Schade, U., Seydel, U., Zähringer, U., Kusumoto, S., Brade, H. — In: Surface structures of microorganisms and their interaction with the mammalian host. Schwartz, U., Richmond, M., Weinheim, Eds. Verlag Chemie (1988).

[3] Lüderitz, O., Staub, A. M., Westphal, O.: Bacteriol. Rev. **30** (1968), 192.

[4] Gaffin, S. L.: Hematology and blood transfusion **29** (1985), 107.

[5] Lachman, E., Pitsoe, S. B., Gaffin, S. L.: Lancet **I** (1984), 981.

[6] Brade, H., Brade, L., Rietschel, E. Th.: Zbl. Bakt. Hyg. A **268** (1988), 151.

[7] Appelmelk, B. J.: Thesis, Free University Press Amsterdam (1987).

[8] Appelmelk, B. J., Verweij-van Vught, A. M. J. J., Maaskant, J. J., Schouten, W. F., Thijs, L. G., MacLaren, D. M.: Antonie van Leeuwenhoek **52** (1986), 537.

[9] Sakulramrung, R., Domingue, G. J.: J. Infect. Dis. **151** (1985), 995.

[10] Siber, G. R., Kania, S. A., Warren, H. S.: J. Infect. Dis. **152** (1985), 954.

[11] BRADE, L., BRADE, H.: Infect. Immun. **50** (1985), 687.
[12] MARKS, M. I., ZIEGLER, E. J., DOUGLAS, H., CORBEIL, L. B., BRAUDE, A. I.: J. Clin. Invest. **69** (1982), 742.
[13] JOHNS, M., SKEHILL, A., McCABE, M. R.: J. Infect. Dis. **147** (1983), 57.
[14] McCABE, W. R., KREGER, B. E., JOHNS, M.: N. Engl. J. Med. **287** (1972), 261.
[15] ZINNER, S. H., McCABE, W. R.: J. Infect. Dis. **133** (1976) 37.
[16] POLLACK, M., HUANG, A. I., PRESCOTT, R. K., YOUNG, L. S., HUNTER, K. W., CRUESS, D. F., TSAI, C. M.: J. Clin. Invest. **72** (1983), 1874.
[17] CLUMECK, N., Delespesse, G., Butzler, J. P. — In: Bacterial endotoxins and host response. Agarwal MK Ed., p. 79. Amsterdam Elsevier/ North Holland Biomedical Press (1980).
[18] ZIEGLER, E. J., McCUTCHAN, J. A., FIERER, J., GLAUSER, M. P., SADOFF, J. C., DOUGLAS, H., BRAUDE, A. I.: N. Engl. J. Med. **307** (1982), 1225.
[19] McCUTCHAN, J. A., WOLF, J. L., ZIEGLER, E. J., BRAUDE, A. I.: Schweiz. Med. Wochenschr. 11, Suppl. **14** (1983), 40.
[20] BAUMGARTNER, J. D., GLAUSER, M. P., McCUTCHAN, J. A., ZIEGLER, E. J., VAN MELLE, G., KLAUBER, M. R., VOGT, M., MUEHLEN, E., LUETHY, R., CHIOLERO, R., GEROUNALOS, S.: Lancet **II** (1985) 59.
[21] CALANDRA, T., GLAUSER, M. P., SCHELLEKENS, J., VERHOEF, J., and the Swiss-Dutch J5 Immunoglobulin Study Group: J. Infect. Dis. **158** (1988) 312.
[22] McCABE, W. R., DEMARIA, Jr. A., BERBERICH, H., JOHNS, M. A.: J. Infect. Dis. **158** (1988) 291.
[23] STEPHAN, W., DICHTELMÜLLER, H., SCHEDEL, I.: Drug. Res. **35** (1985), 933.
[24] APPELMELK, B. J., VERWEIJ-VAN VUGHT, A. M. J. J., MAASKANT, J. J., SCHOUTEN, W. F., THIJS, L. G., MACLAREN, D. M.: FEMS Microbiol. Lett. **40** (1987), 71.
[25] KAÇA, W., DE JONGH-LEUVENINK, J., ZÄHRINGER, U., RIETSCHEL, E. Th., BRADE, H., VERHOEF, J., SINNWELL, V.: Carbohydr. Res. **179** (1988), 288.
[26] TACKEN, A., RIETSCHEL, E. T., BRADE, H.: Carbohydr. Res. **149** (1986), 279.
[27] KOSMA, P., GASS, J., SCHULZ, G., CHRISTIAN, R., UNGER, F. M.: Carbohydr. Res. **167** (1987), 39.
[28] PAULSEN, H., SCHÜLLER, M.: Liebigs Ann. Chem. **247** (1987).
[29] IMOTO, M., KUSUNOSE, N., MATSUURA, Y., KUSUMOTO, S., SHIBA, T.: Tetrahedron. Lett. **28** (1988), 6277.
[30] APPELMELK, B. J., VERWEIJ-VAN VUGHT, A. M. J. J., BRADE, H., MAASKANT, J. J., SCHOUTEN, W. F., THIJS, L. G., MACLAREN, D. M.: Microbial Pathogenesis **5** (1988), 251.
[31] BRADE, H., BRADE, L., NANA, F. E.: Proc. Natl. Acad. Sci. USA, **84** (1987), 2508.
[32] BRADE, L., KOSMA, P., APPELMELK, B. J., PAULSEN, H., BRADE, H.: Infect. Immun. **55** (1987), 462.
[33] ROZALSKI, A., BRADE, L., BRADE, H., KOSMA, P., APPELMELK, B. J., KUSOMOTO, S., PAULSEN, H.: Carbohydr. Res. (in press).
[34] APPELMELK, B. J., VERWEIJ-VAN VUGHT, A. M. J. J., MAASKANT, J. J., SCHOUTEN, W. F., THIJS, L. G., MACLAREN, D. M.: Microbial Pathogenesis **5** (1988), 297.
[35] KIRKLAND, T., ZIEGLER, E. J., TOBIAS, P., WARD, D. C., MICHALEK, S. M., McGHEE, J. R., MACHER, I., URAYAMA, K., APPELMELK, B. J.: J. Immunol. **141** (1988), 3208.
[36] APPELMELK, B. J., COHEN, J., SILVA, A., VERWEIJ-VAN VUGHT, A. M. J. J., BRADE, H., SCHOUTEN, W. F., MOL, O., HONIG, A., THIJS, L. G., MACLAREN, D. M. — In: Proceedings of the 1988 International Symposium on endotoxin, Plenum Press, (in press).
[37] TENG, H. N., KAPLAN, H. S., HEBERT, J. M., MOORE, C., DOUGLAS, H., WUNDERLICH, A., BRAUDE, A. I.: Proc. Natl. Acad. Sci. USA. **82** (1985), 1790.
[38] ZIEGLER, E. J., TENG, N. N. H., DOUGLAS, H., WUNDERLICH, A., BERGER, H. J., BOLMER, S. D.: Clin. Res. **35** (1987), 619 A.
[39] FEELEY, T. W., MINTY, B. D., SCUDDER, C. M., JONES, J. G., ROYSTON, D., TENG, N. H.: Am. Rev. Respir. Dis. **135** (1987), 665.
[40] POLLACK, M., RAUBITCHEK, A. A., LARRICK, J. W.: J. Clin. Invest. **79** (1987), 1421.
[41] LARRICK, J. W., GRAY, O., DEINHART, T., WANG, J., FLATGAARD, J., CHONG, K. T. — In: Bacterial Endotoxins: pathophysiological effects, clinical signifiance, and pharmaceutical control. LEVIN, J., BÜLLER, H., TEN CATE, J. W., DEVENTER, S. J. H., STURK, A., Eds. p. 383. New York. Alan R. LISS, Inc., (1988).
[42] TAUBER, M. G., SHIBL, A. M., HACKBARTH, C. J., LARRICK, J. W., SANDE, M. A.: J. Infect. Dis. **156** (1988), 456.

[43] YOUNG, L. S.: Clin. Res. 32 (1984), 518 A.
[44] HARKONEN, S., SCANNON, P., MISCHAK, R. P., SPITLER, L. E., FOXALL, C., KENNEDY, D., GREENBERG, R.: Antimicrob. Agents Chemother. 32 (1988), 710.
[45] POLLACK, A.: The New York Times, sept 20 (1988).
[46] CHONG, K.-T., HUSTON, M.: J. Infect. Dis. 156 (1987), 713.
[47] WOODS, J. P., BLACK, J. R., BARRITT, D. S., CONNELL, T. D., CANNON, J. G.: Infect. Immun. 55 (1987), 1927.

Book Review

A. FIECHTER (Ed.)

Bioprocesses Including Animal Cell Culture

Berlin: Akademie-Verlag, DDR, 1988. 168 S., 41 Fig., 12 Tab., M. 148,—, ISBN 3-05-500569-4

Der Einsatz tierischer Zellen bei der Gewinnung bioaktiver Substanzen wie Hormonen, monoklonalen Antikörpern und weiteren diagnostisch und therapeutisch wertvollen Proteinen durch Fermentationsprozesse wird zunehmend zu einer wichtigen biotechnologischen Spezialdisziplin, die allerdings in den Grundzügen den Sterilfermentationsprozessen mit Bakterien und Hefen ähnelt. Wichtige Innovationen berücksichtigen die speziellen Erfordernisse, um tierische Zellen im technischen Maßstab zur Vermehrung und Produktion zu bringen.
Dem Herausgeber des Buches ist es gelungen, solche Schwerpunkte in Form von drei Beiträgen unterschiedlicher Autoren (aus Australien, Schottland und Italien) deutlich zu machen:
So wird die Verfahrensentwicklung zur Produktion monoklonaler Antikörper unter dem Aspekt diskutiert, die für mikrobielle Fermentationen auf Basis bekannter metabolischer Pathways, kinetischer Daten und experimentell ermittelter Massenbilanzen ausgearbeiteten Prozeßoptimierungsstrategien analog auf die tierische Zellkultur zu übertragen. Des weiteren wird die Simplifizierung von Nährmedien am Beispiel der Substitution von FKS durch eine Human-Plasmaprotein-Fraktion (Nebenprodukt der Routine-Fraktionierung) vorgestellt und die Produktion von β-Interferon mit humanen Fibroblasten im technischen Maßstab beschrieben.
Ein zweiter Komplex über moderne Methoden der Gentechnik wird durch zwei Arbeiten aus der BRD und Österreich im Sammelband vorgestellt.
Dabei werden Möglichkeiten und Methoden der Oligonukleotid-Chemie für die Konstruktion synthetischer Gene sowie deren Kombination mit Doppelstrang-Genfragmenten und Genen behandelt sowie Clonierungssysteme für Mikroorganismen wie Bakterien, Hefen, filamentöse Pilze und Algen für die Stammverbesserung industriell bedeutender Spezies mittels DNA-Rekombinantentechnik präsentiert.
Das vorliegende Buch kann allen auf den Gebieten der Zellkulturtechnik und der Mikrobenbiotechnologie arbeitenden Biochemikern, Mikrobiologen, Chemikern, Genetikern und Verfahrenstechnikern zum Studium empfohlen werden.

W. BECHSTEDT

Acta Biotechnol. 10 (1990) 1, 23–34 Akademie-Verlag Berlin

The Principle of Optimality as the Foundation of Biotechnological Modeling

PRAUSE, M., KAISER, J.

Academy of Sciences of the G.D.R.
Institute of Biotechnology
Permoserstr. 15, Leipzig 7050, G.D.R.

Summary

There are many points of contact between optimization problems and modeling. On the one hand the model adjustment process itself as a process of estimation is closely connected with optimization, in that it is to produce what is in one sense the best possible model. The basic structure of the optimization problem as problem in decision making with the necessary input of an objective function is thus evident. On the other hand a model is never an end in itself but on the basis of its simulation capacity a means to an end, for example in biotechnological optimization. From this point of view the model is a product of scientific work and thus an economic value. Equally, through its intended purpose the model exhibits a utility value. A complete evaluation of the model as a condition of rational modeling must take into account both these aspects. That is possible in principle by adding the modeling expenditure to expenditure for the realization of biotechnological processes, expressing the economic consequences of model quality as an objective function, and minimizing the specific total expenditure for the product to be produced.
Biotechnological practice requires that the "optimum" model is approached by means of iterative processes. Some practical examples will make the process clear, taking into account qualitative (semantic) and quantitative (accuracy) aspects of the utility value.

Let us begin with an observation relevant to the customary modeling procedure: There are at least two points of contact between mathematical model and optimization. The mathematical model is a) the solution to a problem in optimization and b) a means for the solution of complex biotechnical optimization problems. An examination of the principle of optimality will show that these two aspects are part of a more comprehensive optimization problem. Let us first consider certain basic points of experimental modeling that are present but usually forgotten in practical work. Take a simple statistical operation, the estimation of an expectation value from a sample. The process is obvious enough: Let $\{x_i\}$, $i = 1, \ldots, n$ be the sample of a random value X in standard distribution. Then the mean value $\bar{x}$ of the sample is (in general) taken as an estimate for the expected value μ of the random value:

$$\hat{\mu} = \bar{x} = \frac{1}{n} \sum_{i=1}^{n} x_i.$$

Why is it precisely the mean value and not, say, the median med $\{x_i\}$, the mean x_i, of the ordered sample?
The answer is that the mean value x is the optimum estimate function (in statistical terms "the most efficient" function) for the expectation value. A decision has evidently

be made in favour of one estimate function from the set of all possible estimate functions, by the formulation and solution of a problem in optimization:

$$\min_{\hat{\mu} \in \{\hat{\mu}\}} D^2(\hat{\mu})$$

where: $$D^2(\hat{\mu}) = E[(\mu - \hat{\mu})^2] = \int_{-\infty}^{\infty} (\mu - \hat{\mu})^2 f_{\hat{\mu}}(y) \, dy$$

with $$f_{\hat{\mu}}(y) = \frac{d}{dy} \int\int_{\hat{\mu}(x_1,\ldots x_n)<y} \ldots \int f(x_1) \ldots f(x_n) \, dx_n \ldots dx_1$$

where: $f(x)$ density function of random value X

$f_{\hat{\mu}}(y)$ density function of random value $\hat{\mu}$

$\{\hat{\mu}\}$ set of all estimate functions for μ.

This complicated variable problem itself need not concern us here. We are only concerned with the basic structure. Like all decision making problems the basic structure comprises, in addition to the formulation of all possible solutions, a decision making criterion, which, for formal reasons must be scalar, as otherwise it would only be possible to find an efficient or compromise set in the sense of multicriterial optimization. In the present case the decision making criterion is the standard deviation of the estimate:

$$D^2(\hat{\mu}) = E[(\mu - \hat{\mu})^2] \to \min.$$

But this leaves a further question open: Why exactly this criterion? A criterion in the form

$$E[|\mu - \mu|] \to \min \quad (1)$$

would seem to be more to the point. According to the statistical theory of decision making, $\bar{x}$ is the best estimate for μ for all meaningful criteria, thus also for (1) [7]. In the present case the decision is for once independent of the choice of criterion.

Another example, concerning the decision for an approximation function: Given a function φ by the m points of Fig. 1, whose approximation function is to be a constant c, find the "best approximation" according to the criterion $\|\varphi - c\| \to \min$. The most

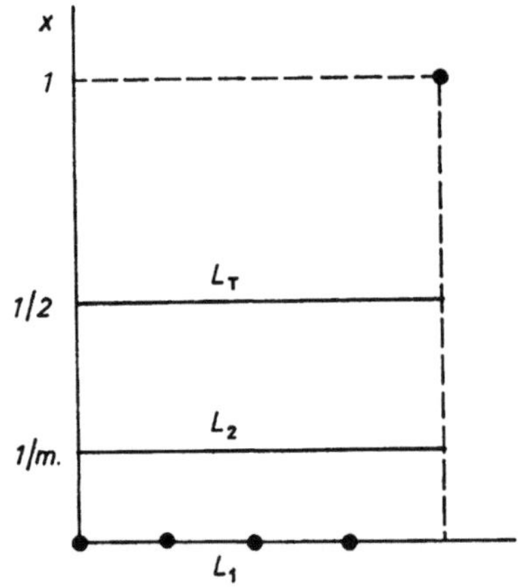

Fig. 1. Approximation of m points x_i by a constant according to L_1-, L_2-, and L_T-norm

frequent norms are:

$$L_1\text{-norm:}\ \|f\|_1 = \sum_{i=1}^{n} |f(x_i)|$$

$$L_2\text{-norm:}\ \|f\|_2 = \sum_{i=1}^{n} |f(x_i)|^2$$

$$L_T\text{-norm:}\ \|f\|_T = \max_{x_i} |f(x_i)|.$$

In the given case, the best approximation is arrived at

by taking L_1-norm: $c = 0$
by taking L_2-norm: $c = 1/m$
by taking L_T-norm: $c = 1/2$

The approximation according to the L_2-norm is nothing other than the GAUSSIAN least squares minimization, the approximation according to the L_T-norm is the CHEBYCHEV-approximation.

It can be seen that the decision for an approximation function takes a very different form according to the optimization criterion adopted. In the given case the difference is to be found above all else in the weight of the "outlier": In the L_1-norm it is of no importance, in the L_2-norm it is of a little importance, and in the case of L_T-norm has a great influence on the decision.

Note that our reservations are not linked to the formal structure of the decision making process, whose requirements have been fulfilled: The optimization problem has been unambiguously formulated by means of a scalar objective function and the inclusion of all possible solutions in the model. The problem is evidently to be found in the content of the objective function and becomes still more weighty if one ask questions of such practical significance as: What aspects of reality should a model deal with? How exact should the model be? It is obviously impossible to answer the question as to the "right" objective function for modeling without going beyond the bounds of the system imposed by the concept of modeling and the necessary experimental data by including the actual purpose of the model [1]. The call for hierarchically superordinated objective functions applies to any decision making process. And therefore to any rational human action.

An especially typical example of this is the work of the technologist. Technology is a purpose orientated scientific discipline, whose end it is always to satisfy human requirements. In general for any one requirement there are a number of technical possibilities. Resources for the implementation of the requirements are on the other hand limited. To deploy the resources to best effect, the technologist must look for technical solutions which call for the least expenditure in fulfilling requirements of society. This requires the constant formulation and solution of optimization problems. To follow these lines is to act on a principle of optimality. We thus regard this principle of optimality as a basic principle in all branches of technology, including biotechnology.

How does the (bio)technologist who is versed in the principle of optimality go about choosing an objective function satisfactory in content? What has he to offer towards rational modeling?

Let us take a typical technological problem. For the satisfaction of certain requirements in society a certain product has to go into production (on an industrial scale). Let us say at this point that in the majority of cases the logic of this step is evident, although it does not spring from the solution of the overall technological problem of optimum distribution of resources.

The technologist's decision respecting the setting up the industrial plant deals with

- aspects of structure (choice of raw materials, microbial production cultures, types and setup of apparatus)
- aspects of design (dimensioning of apparatus)
- aspects of control (determination of the process inputs).

In the following we will include all these desision variables in the vector of optimization variables x.
How does the technologist determine an optimization criterion? With an eye on the general problem of distribution of resources, he relates the total expenditure for the setting up and running of industrial plant to the required output. This serves to formulate the optimization criterion in terms of monetary unit per mass unit of product. This criterion satisfies the formal and actual requirements. It is scalar and produces as a result of optimization the least possible expenditure for the required product [2].
The technological optimization problem is:

$$z_1 = z_1(x, p) \rightarrow \min \qquad (2)$$

where: z_1 specific expenditure [monetary unit/mass unit of product]
x vector of decision variables
p vector of economic and process parameters.

We will not touch on the practical problems of the construction of such an objective function but merely remark that because of their sometimes considerable extent the mathematical formulation of such problems should be pursued diligently and not abandoned too soon. The same may be said in reply to the objection that economic criteria of this type frequently do not encompass all the social requirements. That is true (enough), but no reason not to follow this procedure, which can only enhance objectivity in decision making.
We must draw attention to two fundamental problems.

1. The objective function mentioned is only suitable when all the technical variants result in the same quality of product, in other words the quality is invariant to the decision variables. The quality of the product is the material basis of its utility value which is decisive for the filling of the need. In the case of variable product quality the minimization of the specific expenditure no longer matches up to the decision problem. Besides, the product quality should be maximized. But there must be one sole scalar criterion. This might be attempted by the evaluation of value (expressed as the total expenditure) and utility value. But this is theoretically and practically unsatisfactory. We would recommend taking the frame of balancing for the industrial process beyond the primary product by considering its eventual use, taking this to the point where a product is found whose quality is again invariant to the decision variables. The specific expenditure for this product can be taken as a meaningful optimization criterion.

We have dealt with problems of this kind in SCP-production. As the ultimate use of microbial protein consists in mixed animal feed component, expenditure is no longer set in relation to the primary product SCP with its variable quality but to the mass unit of meat (see Fig. 2) [3].

2. We may note here, that we have an example of the second aspect of the relationship between principle of optimality and modeling referred to above. Complex technological optimum decision making of this kind would be impossible without mathematical models.
This brings us back to the problem of "model quality". The biotechnologist seeks to obtain models of the best possible quality, rightly seeing that the better the model

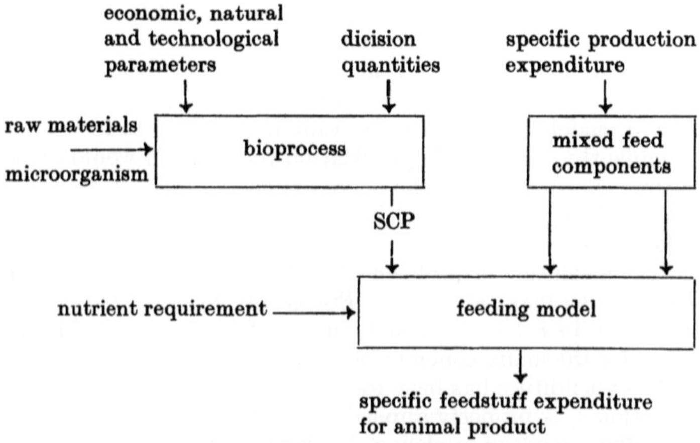

Fig. 2. Optimization of SCP process to the final product criterion

quality (one might also say, the better the utility value of the model) the better the decision making. That is the calculations will more closely tally with the practical results from the plant, and possibly with a reduction of the calculated optimum specific expenditure. Not later than when one sets the value of the model in the objective function as capital outlay, it becomes evident that not the best possible model but the optimum model, that is a model comprising the necessary quality in the sense of the objective function must be aimed at. In short, since higher model quality in general requires an increase in model expenditure, an optimum compromise between value and utility value of the model is required. We discussed an analogous case in respect of product quality. Here the frame of balancing is sufficient. It remains to relate the objective function z_1 to model quality and model expenditure, to enter the relationship model quality/model expenditure in the model, and to take the model expenditure as an additional decision variable.

$$z_2\{x[q(e)], p[q(e)], e\} = z_1\{x[q(e)], p[q(e)]\} + e/A$$

where: z_2 decision criterion specific expenditure including expenditure for modeling
z_1 decision criterion according to (2)
x vector of decision variables
p vector of economic and process parameters
q model quality
e expenditure for modeling
A product output.

Criterion of an optimum decision is thus

$$\min_{e} \left(\min_{x[q(e)]} z_2\{x[q(e)], p[q(e)], e\} \right). \tag{3}$$

This formulation is adapted to two aspects of model quality, which will later appear in two practical examples: In the case of stochastic parameter uncertainty, stochastic decision theory states that with increasing model quality, expressed as reduction in dispersion of model parameters $D^2(P)$, there is a constant improvement in the practical value (that is the value achieved in practice) z_1^*.

The other aspect of model quality is the number of elements of the decision vector. Here the formulation $x[q(e)]$ indicates that with increased expenditure of modeling the quality of the model is improved in the sense that the process is rendered structurally more comprehensive and more decision variables enter into the optimization problem. Here too, with increasing model expenditure, constant improvements in the sense of z_1 are achieved. In principle the solution of optimization problem (3) would answer the questions which have arisen in respect of the necessary model quality. It would supply not only optimum plant design and control but also the model of optimum quality with required model expenditure.

But we see also that a solution of problem (3) in this closed form neither is practically possible, nor does it even make any sense. Practically not possible because it remains impossible to obtain all the necessary models (such as for the relationship model quality/model expenditure) for the many concrete cases. It makes no sense, because the possible conclusion that the expenditure has been too great serves no practical purpose.

The objective function z_2 is properly applied when in iterative form it is used in "real time" according to the process of modeling. In this case it can be used in the evaluation of expenditure of any kind in research and development and forms the basis for the control of R & D according to the requirements set by society.

In keeping with the complexity of this problem it is advisable to isolate the various practical steps; first, as early as possible in the process, concentrating on the estimating of the utility in z_1 that can be expected, and to establish further expenditure on the basis of this utility. This procedure will not in all concrete cases be taken to the point of the approximated optimum expenditure. The special virtue of the method consists in being able to recognize factors of no great value and exclude them at an early stage [4].

Let us now take two examples from industrial microbiology. The first deals with the question of what aspects of reality should be considered in modeling, more particularly, whether it is worth while extending the model to include an additional control variable for optimization. First let it be said that an optimization including a model extended by decision vector v can make very good sense because in contrast to simple optimization, it supplies equally good or better results in the optimum optimization criterion:

$$z_1^*(x^*) = \min_x [z_1(x)] \geqq \min_{x,v} [z_1(x, v)] = z_1^{**}(x^{**}, v^*).$$

But this still does not tell us if such a procedure will pay off. For this we must compare the proposed improvement in z_1^* with the necessary expenditure.

Let the concrete situation be as follows: In the optimum design of continuous aerobic fermentation process, the reaction temperature ϑ can in the initial stages be given a fixed value, instead of being taken as an optimization variable. This value should be the "biological optimum", that is, the temperature at which the specific rate of growth is at a maximum, and according to the theory of Pirt, the α-values are at a minimum [5]:

$$\vartheta^0 = (\vartheta \mid \alpha_x^{\,i} = \alpha_x^{\,i_{\min}} = \alpha_x^{\,i0}).$$

This value is usually known approximately from preparatory microbial investigations for industrial cultures. It is also known, that the temperature dependence of $\alpha_x^{\,i}$ of microorganisms takes the schematic form in reduced (dimensionless) coordinates seen in Fig. 3. This makes it clear that $\vartheta < \vartheta^0$ can not be used for the optimum design, since with increased $\alpha_x^{\,i}$ the process economy would decrease as a result of increased energy expenditure for oxygen transfer and substrate and cooling expenditure. A reduction in temperature itself would produce a similar result. It would mean a reduction in the driving force for the necessary heat transfer and thus again an increase in cooling expenditure. The same applies in the case of $\vartheta > \vartheta^0$ for α-values, but as a result of the reduction

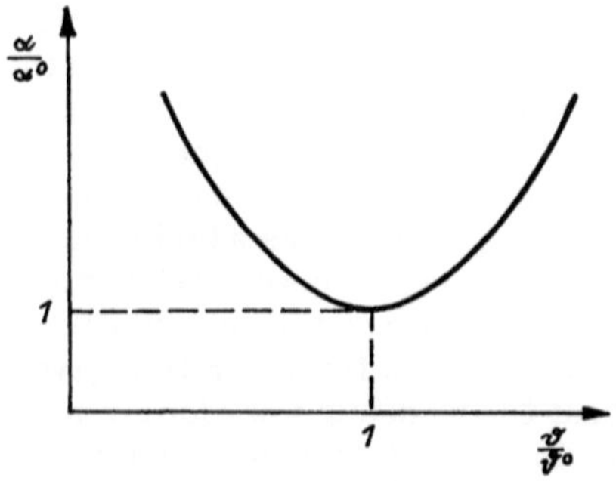

Fig. 3. Dependence of specific demand α_x^i on temperature ϑ in dimensionless coordinates, with ϑ^0, $\alpha_x^{i,0}$ as the biological optimum

in cooling expenditure together with increasing driving force a compromise situation arises. As a result the optimum reaction temperature ϑ^* of z_1 is equal to or greater than the biological optimum:

$$\vartheta^* = (\vartheta \mid z_1 = z_1^*) > \vartheta^0$$

and thus always

$$z_1^*(x_1^*, \vartheta^0) = \min_x z_1(x, \vartheta^0) > \min_{x,\vartheta} z_1(x, \vartheta) = z_1^{**}(x^{**}, \vartheta^*).$$

But before we proceed to find a solution for the pertinent optimization problem and set up a model for the dependence $\alpha_x^i = \alpha_x^i(\vartheta)$ (as we in the meantime know, for $\vartheta > \vartheta^0$ only), a procedure not without metrical difficulties (we will return to this point in the second example), it is advisable to be clear about the probable utility. This is shown in the graph in Fig. 4. For fixed and arbitrary values ϑ and α_x^i (here represented by oxygen demand $\alpha_x^{O_2}$) the optimization problem is solved in respect of z_1. Fig. 4 shows the result in

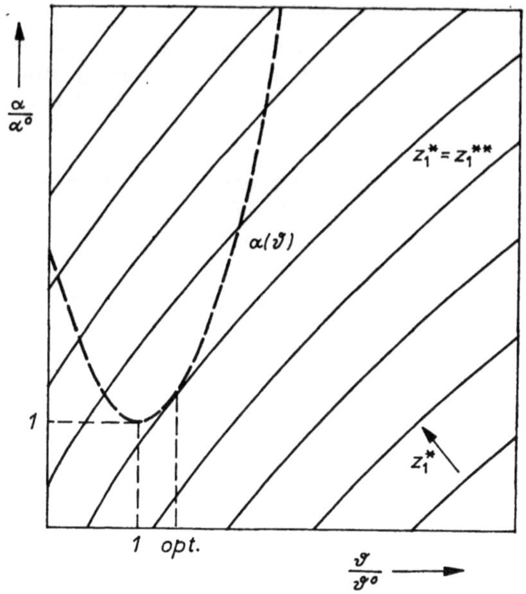

Fig. 4. Finding of optimum temperature ϑ^* according to objective function z_1 depending on ϑ and $\alpha_x^{O_2}$, and $\alpha_x^{O_2}$ depending on ϑ

reduced coordinates for ϑ and $\alpha_x^{O_2}$ with z_1^*, in the form of level lines. The graph shows a hypothetical model $\alpha_x^{O_2} = \alpha_x^{O_2}(\vartheta)$ as a dotted line. If optimization with the temperature as an additional decision variable is to produce a better economic result, this curve must intersect with the level lines at $\vartheta/\vartheta^0 = 1$. This shows the possibility of obtaining a first estimate of the probable utility with few initial items of information on the dependence $\alpha_x^{O_2}(\vartheta)$, in an extreme case two tuple only, or in doubtful cases finding a suitable starting point for further experiments or breaking off evidently unsuccessful experimentation.

This is the course pursued (by us) in the optimization of industrial SCP-production. When hypothetical models on the basis of data from literature gave promise of an optimization up to 5% of z_1^*, we set about the modeling work for given microorganism/substrate combinations. Depending on the actual economic and process conditions improvements of 2–3% of z_1^* for $\vartheta^* > \vartheta^0$ were achieved.

Such an improvement justified the modeling expenditure throughout the process.

Now to the second example. Here we must address ourselves to a second aspect of model quality, the accuracy of the model. As we already pointed out we can utilize the scatter $D^2(P)$ of the model parameters as the measure of accuracy. The majority of technological decisions work with models that, on the basis of stochastic parameters, exhibit a non-measurable deviation from reality. In such a case, a technological decision on the basis of an unequivocally forecastable utility is not possible. The theory of decision making under uncertainty and risk [9] supplies the appropriate means for decision making by formulating the (scalar) "preferential functions". Of particular value in technological optimization problems under uncertainty is the "decision by the BAYES criterion" with the preferential function

$$k(x) = E[Z_1(x, P)] = \int_p \cdots \int z_1(x, p)\, \mathrm{d}F,$$

where:
E expected value
F distribution function of the stochastic parameter P
x vector of decision variables
P vector of stochastic parameters.

In biotechnology this "stochastic optimization" procedure is hardly ever used nowadays, because of the considerable amount of calculation involved among other reason. The problem is considerably simplified if $z_1(x, p)$ is a linear function of p. Then and only then

$$k(x) = z_1(x, E(P)),$$

that is to say, optimization of deterministic models employing the expectation value of P provides the decision by the BAYES criterion. The linearity of p can frequently presumed if the scatter of P is sufficiently small (as in this example).

We now come to the question as to the accuracy (the standard deviation of the parameters) that is to be expected of the model. For this it is necessary to know, with decision by the BAYES criterion x^*, how far the actual minimum of the objective function z_1 will probably not be achieved, depending on $D^2(P)$. This discrepancy is expressed in the loss function l:

$$l(x, p) = z_1(x, p) - \min_x z_1(x, p).$$

In particular it is of practical interest to know the "mean loss" or "mean risk":

$$E[L(x, P)].$$

(We note here that a decision x^* by the BAYES criterion also supplies minimum of the mean loss:

$$E[L(x^*, P)] = \min_x E[L(x, P)]).$$

Here some explanations on the mean loss, using one (real) decision variable x and one (real) parameter p, where the objective function z_1 depends linearly on p and is differentiabel.
Fig. 5 is a schematic representation of the dependence of the objective function z_1 on the decision variable x and the parameter p. The envelope of the family of lines supplies the minimum objective function values for fixed p. The loss function is the interval between the envelope and the chosen straight line $x =$ const.

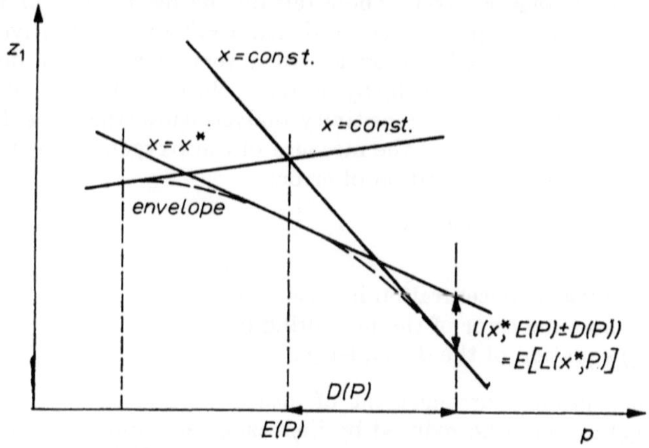

Fig. 5. Objective function z_1 depending on decision x and parameter p, and expected loss l of an optimum decision x^* depending on $D(P)$

The optimum decision x^* by the BAYES criterion is by definition the deterministic optimum for the expectation value of the parameter P. The loss function $l(x^*, p)$ may be represented approximately by the objective function $z_1(x, p)$ as TAYLOR series up to the quadratic term:

$$\begin{aligned} l(x^*, p) &= z_1(x^*, p) - \min_x z_1(x, p) \\ &= z_1(x^*, p) - \{z_1(x^*, E(P)) + \partial z_1/\partial p\,[p - E(P)] + \partial z_1/\partial x\, \Delta x \\ &\quad + 1/2\, \partial^2 z_1/\partial x^2\, (\Delta x)^2 + \partial^2 z_1/\partial x\, \partial p\, \Delta x\, [p - E(P)] \\ &\quad + 1/2\, \partial^2 z_1/\partial p^2\, [p - E(P)]^2\}. \end{aligned}$$

Resulting from the condition that along the envelope $\partial z_1/\partial x = 0$ it follows that

$$\partial z_1/\partial x \approx \partial^2 z_1/\partial x^2\, \Delta x + \partial^2 z_1/\partial x\, \partial p [p - E(P)] = 0,$$

thus

$$\Delta x = -(\partial^2 z_1/\partial x\, \partial p)/(\partial^2 z_1/\partial x^2),$$

and thus

$$l(x^*, p) = [(\partial^2 z_1/\partial x\, \partial p)^2/(2\, \partial^2 z_1/\partial x^2)]\, [p - E(P)]^2. \tag{4}$$

As mean loss results

$$\begin{aligned} E[L(x^*, P)] &= \int_P l(x^*, p)\, f(p)\, \mathrm{d}p \\ &= [(\partial^2 z_1/\partial x\, \partial p)^2/(2 \partial^2 z_1/\partial x^2)] \int_P [p - E(P)]^2\, f(p)\, \mathrm{d}p \end{aligned}$$

$$E[L(x^*, P)] = [(\partial^2 z_1/\partial x\, \partial p)^2/(2 \partial^2 z_1/\partial x^2)]\, D^2(P). \tag{5}$$

Because of (4) this expectation value is nothing other than the value of the loss function for $p = E(P) \pm D(P)$:

$$E[L(x^*, P)] = l(x^*, E(P) \pm D(P))$$

with the result that the mean loss in Fig. 5 can be read off immediately.
Equation (5) express the relation sought between the degree of uncertainty of the model and the expected loss. Let us give our example in concrete terms. The output for the design of a fermentation stage is given. There remain the reactor size and the energy input as decision variables. The specific oxygen demand $\alpha_x^{O_2}$ plays a decisive part in the vector of the model parameters. z_1^* and x^* are to a very great extent dependent on this parameter. Arriving at $\alpha_x^{O_2}$ practically by means of the oxygen and biomass balance is fraught with considerable error [6]. Sensitivity analysis shows that it is dependent to a very large degree on the accuracy of the measure of the oxygen concentration in the gas phase. On the principle of propagation of error:

$$D(\alpha_x^{O_2}) = \mathrm{d}\alpha_x^{O_2}/\mathrm{d}\psi_{O_2}\, D(\psi_{O_2})$$

where

ψ_{O_2} oxygen concentration in waste gas
$D(\psi_{O_2})$ nominal error of the measuring instrument
$\alpha_x^{O_2}(\psi_{O_2})$ equation of the dependence.

Having selected the optimum arrangement of measuring and the optimum range of measuring [6], $D(\psi_{O_2})$ can only be reduced by increasing the number of experiments or by refinement of the accuracy of the measuring instrument. Both involve expenditure. Let us examine the latter case. For a clear view of the expenditure/utility ratio it is best, to examine the reduction of the loss function in respect of a point of reference. Since the reduction of the loss function is synonymus with an increase in the expected utility it is possible to formulate a utility function

$$u(D) = l(x^*, E(P) \pm D_0) - l(x^*, E(P) \pm D),$$

where D is the parameter accuracy (in this case of $\alpha_x^{O_2}$) which must be compared with the point of reference D_0.
This must be compared with expenditure $e(D)$ which is necessary to arrive at accuracy D.
It is not our intention to take our example to the point of estimating the optimum accuracy D^*. We see from Fig. 6 that each and every procedure to improve accuracy makes

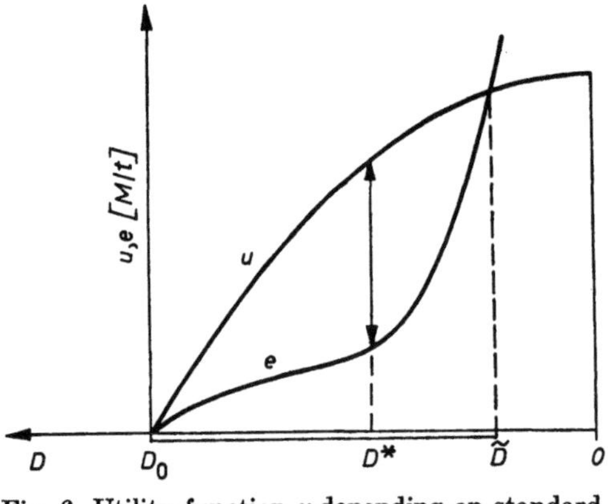

Fig. 6. Utility function u depending on standard deviation D of the stochastic parameter, and expenditure e for reducing D

economic sense to the point $\tilde{D}$, since throughout the entire range D_0 to $\tilde{D}$ utility exceeds expenditure. As will usually be the case, in our particular situation ideas about the relationship expenditure/D-reduction were rather hazy. In such a case it is a practical procedure to give the utility function as upper limit for expenditure e to reduce the measurement error from D_0 to D.
In our particular case we arrived at the utility function

$$u = 98.424 - 10.936(\Delta\psi_{O_2}/\psi_{O_2})^2$$

(Fig. 7). For paramagnetic ψ_{O_2}-measurement we used an instrument with a practical accuracy of 3%. The nominal accuracy of 2% could only be achieved by means of an increase in maintenance expenditure, in particular for the preparation of the gas to be measured.

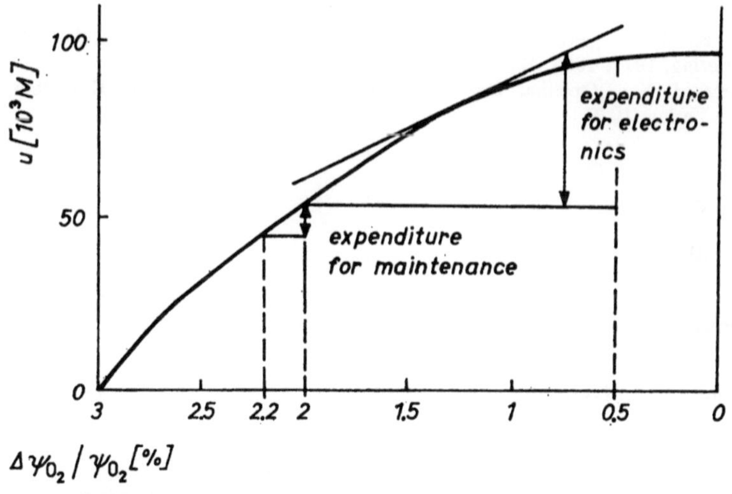

Fig. 7. Utility u of reducing measuring error of O_2-concentration, $\Delta\psi_{O_2}/\psi_1$ and expenditure for two ways of reducing $\Delta\psi_{O_2}/\psi_{O_2}$

Was this expenditure really worth while?
According to Fig. 7 it was: The maintenance expenditure is always justified when the initial error is greater than 2.2%. Of further interest was the question as to whether an increase in nominal accuracy of the meassuring instrument could be economically achieved by improvement in the electronics of the instrument. For this purpose a linear relationship between expenditure/accuracy (see Fig. 7) was estimated, to be valid to approximately $\Delta\psi_{O_2}/\psi_{O_2} = 0.5\%$. Fig. 7 shows that in this way there is economic advantage only for increased accuracy to approximately $\Delta\psi_{O_2}/\psi_{O_2} = 1.2-1.4\%$.
It can be seen from Fig. 7 that the possibility of devising a measuring instrument working on other (measuring) principles for measurement error of less than 0.5%, entails no economic advantage in this case: The expected utility is bound to be small, whereas the expenditure is bound to be considerably higher.

Received November 1, 1988

References

[1] CYPKIN, Ya. Z.: "Osnovy teorii identifikacii", izd. Nauka, Moskva 1984.
[2] HILDEBRAND, H.-J., HEDRICH, P., UFER, D.: „Wirtschaftlichkeitsrechnung", VEB Verlag für Grundstoffindustrie, Leipzig 1970.
[3] SOYEZ, K., PRAUSE, M., PÖHLAND, D. — In: „Entwicklungsprobleme und Effektivitätsermittlung in der Biotechnologie", Institut für Theorie, Geschichte und Organisation der Wissenschaft, AdW d. DDR, Kolloquiumsbericht 36 (1987), 79.
[4] PRAUSE, M., PÖHLAND, D., RINGPFEIL, M. — In: „Informationen für die Leitung der Forschung", AdW d. DDR, wissenschaftliches Informationszentrum Berlin, GW 29 (1984), 66.
[5] PRAUSE, M., SOYEZ, K.: Acta Biotechnol. 2 (1982), 73.
[6] SOYEZ, K., THEISS, D., PRAUSE, M.: Abh. d. AdW d. DDR, N 3 (1979), 299.
[7] EBERL, W., MOESCHLIN, O.: „Mathematische Statistik", Berlin, New York: DE GRUYTER, 1982.
[8] HUMAK, K. M. S.: „Statistische Methoden der Modellbildung", Bde. I—III, Berlin: Akademie-Verlag, 1977, 1983, 1984.
[9] SENGUPTA, J. K.: „Optimal Decision under Uncertainty", Berlin, Heidelberg, New York: Springer-Verlag, 1979.

Book Review

T. E. POPOVA

Entwicklung der Biotechnologie in der UdSSR

Moskau: Verlag Nauka, 1988, 200 S., 6 Abb.

Herausbildung und Entwicklungsbedingungen der Biotechnologie als neues komplexes Wissenschaftsgebiet zu kennzeichnen, u. a. als eine Grundlage für Wissenschaftsleitung und Strategiebildung, ist das Anliegen der Autorin. Ausgehend von einer Begriffsbestimmung und den Besonderheiten der Biotechnologie werden ihre Komplexität, darunter ihr Zusammenhang mit biotechnologischer Industrie, dargestellt und vier Entwicklungsphasen hervorgehoben: empirische Nutzung biologischer Organismen und Leistungen, Entstehung grundlegender biologischer Wissenschaft, Schaffung von Wissenschaft und Praxis mikrobiologischer Industrie, Herausbildung der wissenschaftlich-technischen Grundlagen für die moderne, gegenwärtige Biotechnologie. Dabei wird besonders auf den Zusammenhang von traditionellen und neuartigen Biotechniken, auf die Entwicklung von Apparaturen und Ausrüstungen für neue Richtungen und auf mikrobielle Biosynthese eingegangen. Unter den traditionellen Gebieten werden Antibiotika-Herstellung, industrielle Eiweißgewinnung, die Produktion von Aminosäuren und Vitaminen, Enzymtechnik und ihre Anwendungsrichtungen sowie biotechnologische Einflüsse auf Umweltgestaltung, Energie- und Rohstoffprobleme näher beschrieben.
Gen- und Zelltechnik, ihre Entstehung und Aussichten, stehen im Mittelpunkt der Ausführungen zu neuen Methoden der Biotechnologie. Grundlegende Anforderungen moderner Biotechnologie und ihrer industriellen Nutzung sowie Entwicklungstendenzen zeigt die Autorin im Hinblick auf die Gewinnung und Vervollkommnung mikrobieller Produzenten, die Enzymtechnologie, die Aufarbeitung, ökologische Erfordernisse, auf Apparaturen und Ausrüstungen sowie die Modellierung und Optimierung biotechnischer Prozesse, abgerundet durch einen Ausblick auf absehbare Tendenzen möglicher biotechnologischer Beiträge zum wirtschaftlichen Fortschritt in der Sowjetunion.

P. HANKE

Acta Biotechnol. 10 (1990) 1, 35—40 Akademie-Verlag Berlin

Production of High Ethanol Concentrations from Glucose Using a Vertical Rotating Immobilized Cell Reactor of the Bacterium Zymomonas mobilis

AMIN, G.[1], DOELLE, H. W.[2]

[1] Department of Microbiology, Facultat Agriculture
Cairo University, Giza, Egypt
[2] Department of Microbiology, University of Queensland
St. Lucia-Brisbane, Qld. 4067, Australia

Summary

Long-term continuous ethanol production of up to 80 $g \cdot l^1$ with a volumetric ethanol productivity of 63 $g \cdot l^{-1} \cdot h^{-1}$ was maintained for more than 72 days using a Vertical Rotating Immobilized Cell Reactor of the bacterium *Z. mobilis*. Continuous production of higher ethanol concentration was unsuccessful due to an inhibition of cell growth by long exposure to high ethanol concentrations. However, ethanol concentration as high as 120 $g \cdot l^{-1}$ and volumetric ethanol productivity of 13 $g \cdot l^{-1} \cdot h^{-1}$ were achieved in a repeated-batch fermentation system using the same bioreactor. By a simple washing operation at the end of each run, immobilized biomass could be effectively regenerated and used to carry out more than 10 successive fermentation cycles.

Introduction

Renewed interest in ethanol as a biofuel, octane enhancer and chemical feedstock has stimulated researchers of different disciplines to investigate methods for increasing rates and yields of fermentation [1, 2, 3]. It is important to obtain high ethanol concentrations in fermentation broth to reduce the distillation costs and high volumetric ethanol productivity to reduce the bioreactor costs [4]. However, conventional fermentation with low concentration of freely suspended cells is associated with extremely low volumetric ethanol productivity, and uses comparatively low substrate concentrations to avoid inhibitory, reactions of cell growth by high glucose concentrations. Recently, immobilized cell bioreactors have been reported to increase both volumetric ethanol productivity and final ethanol concentration [5]. The packed-bed bioreactor was the most frequently used type with immobilized cells [6, 7, 8]. However, such bioreactor suffers a serious probleme of unstability on long-term operation. Due to continuous biomass accumulation and CO_2 holdup, significant mass transfer limitations both for ethanol and fermentative substrate out and into cell mass and CO_2 channelling problems are introduced in such bioreactor causing the overall fermentation rate and productivity to decline [9, 10]. In a previous study [11] a Vertical Rotating Immobilized Cell Reactor (VRICR) with a constant immobilized cell concentration was developed and used successfully for continuous glucose conversion into ethanol. The objective of this work was to examine the feasibility of using such bioreactor to convert more concentrated glucose solution into higher ethanol concentration in order to reduce the distillation costs.

Materials and Methods

Microorganisms and Culture Media

Zymomonas mobilis UQM 2716 (ATCC 396760) was used. It was maintained on a stock culture at 6 °C in a liquid medium containing 2.0 g · l^{-1} of casein hydrolysate, KH_2PO_4 and $(NH_4)_2SO_4$; 0.05 g · l^{-1} $MgSO_4 \cdot 7H_2O$; 5.0 g · l of yeast extract and 50 g · l^{-1} of glucose. Glucose was sterilized separately from the other medium components and mixed aseptically. The same medium was used for the fermentation experiments but with different concentrations of glucose.

Bioreactor

The Vertical Rotating Immobilized Cell Reactor (VRICR) described in an earlier study [11] was used. *Z. mobilis* cells were entrapped into polyurethane foam (97% porosity) discs sandwiched between two perforated metal sheets to form an Immobilized Biomass Unit (IBU). A sum of 10 IBUs were fixed along a rotating shaft, within the bioreactor, in two separate stacks. The IBUs were rotated at a constant speed of 6 rpm throughout the whole fermentation experiments. The total volume of the bioreactor was 3.0 litres.

Immobilization Technique

The technique previously described [12], was used to entrap *Z. mobilis* cells into polyurethane foams.

Analytical Methods

Glucose was estimated using a YSl Model 27 Industrial Analyzer. Ethanol was determined by gas chromatography (Varian 3400) with Poropak Q as column material, nitrogen as carrier gas (75 ml · min^{-1}), and column temperature of 200 °C. The efficiency of glucose conversion into ethanol was calculated as the percentage of the theoretical value of 2 moles of ethanol per mole of the initial glucose concentration.

Biomass Concentration

The cell concentration in polyurethane foams (immobilized biomass) or in the culture broth (freely suspended cells) were followed by estimating the dry weight of the cell mass after drying at 105 °C and desiccation to a constant weight. To obtain the immobilized biomass concentration, samples of polyurethane foams were taken, squeezed several times, shaken and washed in physiological saline solution for 10 min. The calculation was based on the total reactor volume.

Cell Viability

The total viable count of the bacterial cells was determined using the standard plate count method [13] and converted into grams of viable biomass. A cell weight of 10^{-12} g · $cell^{-1}$ was assumed [8]. Cell viability is the percentage of viable biomass.

Results and Discussion

The conversion of high concentrations of glucose into ethanol was investigated using VRICR which was prepared as described in Materials and Methods section. The medium containing 160 or 200 g · l^{-1} of glucose was pumped into the reactor at dilution rate of 0.7 h^{-1}. Samples were taken and analyzed for immobilized biomass, cell viability, residual glucose and ethanol. Fig. 1 shows the results. A steady state with no residual glucose was reached only when a medium containing 160 g · l^{-1} of glucose was fed into the

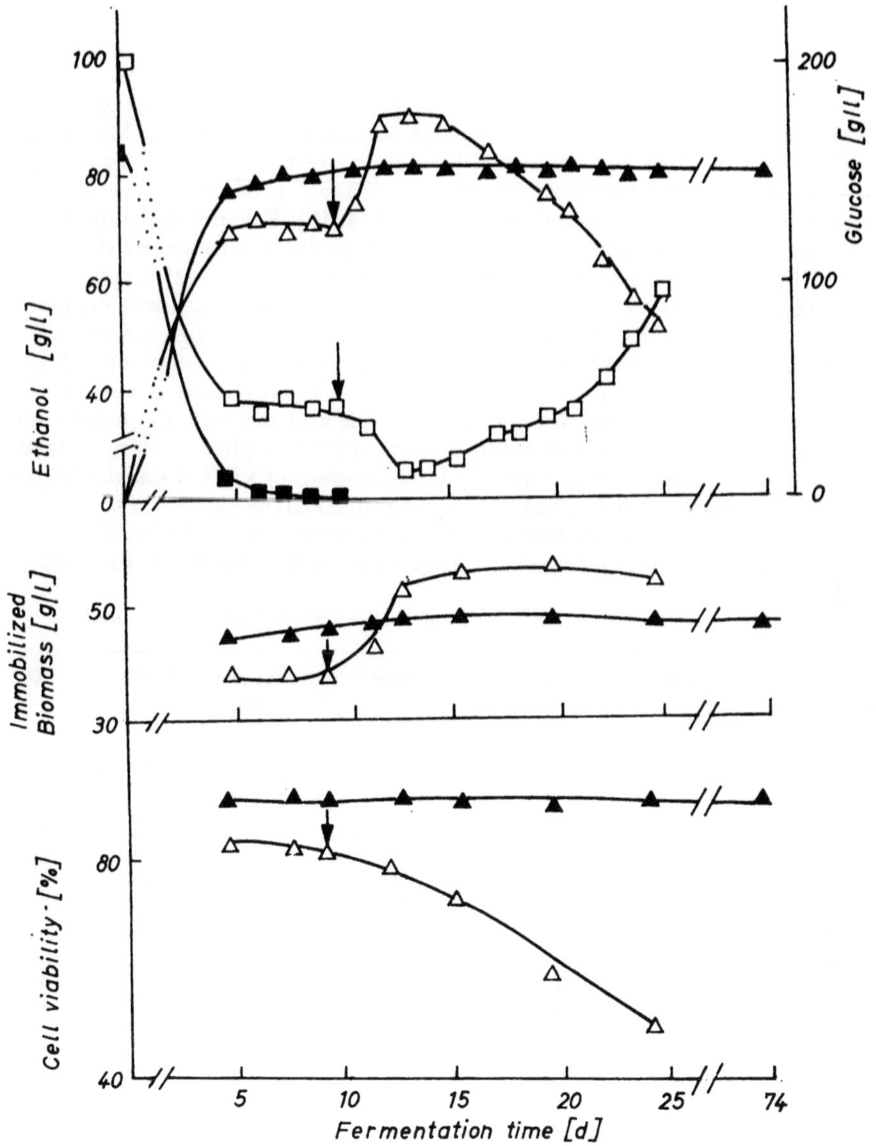

Fig. 1. Performance of VRICR of the bacterium *Zymomonas mobilis* during continouus conversion of 160 (——▲——) and 200 (——△——) g · l⁻¹ of glucose into ethanol. The arrow indicates the change in dilution rate from 0.7 to 0.5 h^{-1} when medium contains 200 g · l^{-1} of glucose was fed into the reactor

bioreactor. Ethanol concentration as high as 80 g · l^{-1} and an immobilized biomass concentration of 42.5 g · l^{-1} with a cell viability of more than 90% were maintained. When a medium containing 200 g · l^{-1} of glucose was fed into the bioreactor, a different performance was noticed. A steady state concentrations of 70 g · l^{-1} of ethanol, 35 g · l^{-1} of immobilized biomass with a cell viability of 80% and up to 43 g · l^{-1} of residual glucose were maintained. However, when the dilution rate was decreased down to 0.5 h^{-1}, better performance was noticed. Ethanol and immobilized biomass concentrations

increased and attained values of 91 and 56 g · l^{-1} resp. The concentration of residual glucose decreased down to 8.0 g · l^{-1}. This new steady state lasted only for 36 hours after which ethanol concentration decreased gradually and reached the lowest value of 50 g · l^{-1} on day 25. Such dramatic drop in reactor performance might be explained by continuous exposure of immobilized biomass to high concentrations of ethanol in the presence of residual glucose. This was confirmed by a sharp drop in cell viability down to 45%. Similarly, unsuccessful long-term continuous ethanol production was reported by other researchers using immobilized cell systems [14, 15]. Ethanol was reported to inhibit both microbial growth and metabolism [16]. It is thought to act by dissolving the membrane lipids, thus affecting the membrane fluidity [17], and increasing its permeability [18]. The resulting efflux of ions, cofactors and coenzymes was suggested to be the primary reason for ethanol inhibition of free suspended cells of *Z. mobilis* [19].
In an attempt to reduce cell growth inhibition by ethanol during conversion of high glucose concentrations, a repeated-batch fermentation system was proposed. A new VRICR prepared and started with the bacterium *Z. mobilis* cells and fed with a complete medium containing 100 g · l^{-1} of glucose at a dilution rate of 0.7 h^{-1}. After reaching a steady state with 25 g · l^{-1} of immobilized biomass, the liquid content of VRICR was discharged and a new medium containing 200 g · l^{-1} of glucose was added to the bioreactor to start the first batch. By determining the residual glucose in the fermentation broth, the termination of the fermentation was established. Tab. 1 shows the results of 10 repeated

Tab. 1. Repeated-batch fermentations using VRICR of the bacterium *Zymomonas mobilis*

Fermentation Run	Initial concentrations [g · l^{-1}]				Final concentrations [g · l^{-1}]				Fermentation time	P*	E**
	Glucose	Ethanol	Biomass		Glucose	Ethanol	Biomass				
			a	b			a	b	[h]	[g · l^{-1} · h^{-1}]	[%]
1	200.7	3.0	22.0	0.74	0.8	100.9	35.9	2.5	9.0	10.9	95.4
2	200.0	10.0	33.4	2.10	30.4	93.7	31.8	2.3	9.0	9.2	81.0
3	201.5	20.0	27.5	2.50	74.2	81.8	28.7	3.0	9.0	6.9	60.0
4	198.7	1.0	9.5	0.40	0.0	97.1	29.1	2.1	9.5	10.1	94.6
5	240.1	1.5	10.3	0.45	1.7	119.3	31.5	2.4	9.0	13.1	96.0
6	243.5	1.2	9.4	0.70	2.1	120.8	30.2	2.8	9.5	12.6	96.1
7	239.8	0.9	8.8	0.60	1.5	117.9	27.7	2.6	9.5	12.3	95.5
8	241.4	1.4	9.8	0.55	1.0	118.4	33.0	2.7	9.0	13.1	94.8
9	242.4	1.6	10.1	0.75	2.0	120.3	34.4	3.0	9.0	13.2	95.8
10	242.7	1.8	11.0	0.57	1.5	120.4	35.1	2.8	9.0	13.3	95.6

* P is volumetric ethanol productivity ** E is the efficiency of glucose conversion into ethanol; a, immobilized biomass; b, free biomass

batch fermentations. After 9 hours, almost 200 g · l^{-1} of glucose were completely converted converted into ethanol with a volumetric ethanol productivity of 10.9 g · l^{-1} · h^{-1}, in the first batch. However, a dramatic drop in glucose conversion into ethanol was noticed in the second and third batch, and up to 72 g · l^{-1} of residual glucose were detected in the fermentation broth. This might be explained by an increased level of undesired metabolites which might have been produced and carried over from previous batches. *Z. mobilis* species are known to produce small quantities of organic products other than ethanol, such as acetaldehyde, higher alcohols, lactic acid and acetoin [3, 20]. One or more of these could inhibit the cells in relatively low concentrations as occurs in *Saccharomyces cerevisiae* [21]. However, such metabolites may act together with initial high glucose concentrations so as to suppress cell growth and hence ethanol production

rates. An initial ethanol concentration, in the range of 0 to 3.0% and in the presence of 100 g · l^{-1} of glucose, was reported to inhibit the growth of *Z. mobilis* cells [22]. Such inhibition was approximately linear in ethanol concentrations.
To solve this problem, the liquid content of VRICR was discharged and the immobilized biomass was washed, by adding sterile saline solution to the bioreactor, and rotated at 12 rpm for 3.0 min, after which saline solution was discharged carrying excess biomass and undesirable metabolites. A new medium containing 200 g · l^{-1} of glucose was then added and the rotation speed returned back to the original rate of 6 rpm. The activity of VRICR was restored and complete glucose conversion into ethanol was achieved in the fourth batch, with no residual glucose.
From now on, immobilized biomass was washed after termination of each batch and before the next. Under such conditions, 6 further fermentation batches of higher glucose concentrations (up to 240 g · l^{-1}) were successfully conducted using the same immobilized biomass. A glucose to ethanol conversion efficiency of more than 95% and volumetric ethanol productivity of up to 13 g · l^{-1} · h^{-1} was reported for repeated batch fermentation system using growing yeast cells immobilized in K-carrageenan gel [14]. Cell entrappment into gels is more likely to be limited by diffusional restrictions than that in relatively open structure of polyurethane foams [5, 12, 23].

Conclusion

VRICR can produce up to 80 g · l^{-1} of ethanol continuously and indefinitly with a volumetric ethanol productivity of 63 g · l^{-1} · h^{-1}. Higher ethanol concentration (up to 120 g · l^{-1}) could also be achieved using such bioreactor in a repeated-batch fermentation system with a volumetric ethanol productivity of 13 g · l^{-1} · h^{-1}. Those values are 10 times higher than volumetric ethanol productivities of the corresponding traditional fermentations using free cells. In addition to its superiority in maintaining stable long-term operation, VRICR provides the most attractive alternative for a batchwise industrial ethanol production system. Only IBUs have to be constructed and fixed along the presently existing shaft which is rotated at a relatively low speed maintaining the energy costs far lower than of free cell fermenter. Moreover, by simple washing operation, immobilized biomass can practically be regenerated and used indefinitely in a repeated-batch fermentation system.

Acknowledgement

We are grateful for the assistance given to us by the University of Queensland through a University of Queensland Postdoctoral Fellowship (G.A.) and University Grant Allowances.

Received December 28, 1988

References

[1] ROGERS, P. L., LEE, K. J., TRIBE, D. E.: Biotechnol. Lett. 1 (1979), 165.
[2] CROMIE, S., DOELLE, H. W.: Eur. J. Appl. Microbiol. Biotechnol. 14 (1982), 69.
[3] AMIN, G., VANDEN EYNDE, E., VERACHTERT, H.: Eur. J. Appl. Microbiol. Biotechnol. 18 (1983), 1.
[4] ROSE, D.: Process Biochem. 12 (1976), 10.
[5] KOLOT, F. B.: Process Biochem. 15 (1980), 2.
[6] WADA, M., KATO, J., CHIBATA, I.: Eur. J. Appl. Microbiol. Biotechnol. 10 (1980), 275.

[7] AMIN, G., VERACHTERT, H.: Eur. J. Appl. Microbiol. Biotechnol. **14** (1982), 59.
[8] ARCURI, E. J.: Biotechnol. Bioeng. **24** (1982), 595.
[9] JAIN, V. K., TORAN-DIAZ, I., BARATTI, J.: Biotechnol. Bioeng. **27** (1985), 273.
[10] SOLA, C., CASAS, C., GODIA, P., POCH, M., SERRA, A.: Biotechnol. Bioeng. Symp. **17** (1986), 519.
[11] AMIN, G., DOELLE, H. W.: Biotechnol. Tech. **3** (1989), 95.
[12] AMIN, G., DOELLE, H. W., GREENFIELD, P. F.: Biotechnol. Lett. **9** (1987), 225.
[13] POSTGATE, J. R. — In: NORRIS, J. R., RIBBONS, D. W. (eds.) Methods in Microbiology, 1969, Vol. 1, Academic Press, London New York.
[14] AGRAWAL, D., JAIN, V. K.: Biotechnol. Lett. **8** (1986), 67.
[15] BAIPAI, P. K., MARGARITIS, A.: J. Ferment. Technol. **65** (1987), 233.
[16] GHOSE, T. K., TYAGI, R. D.: Biotechnol. Bioeng. **21** (1979), 1387.
[17] ROWE, E. S.: Biochemistry **22** (1983), 3299.
[18] DOMBEK, K. M., INGRAM, L. O.: J. Bacteriol. **157** (1984), 233.
[19] OSMAN, Y. A., INGRAM, L. O.: J. Bacteriol. **164** (1985), 173.
[20] SWINGS, J., DE LEY, J.: Bact. Rev. **41** (1977), 1.
[21] MARIORELLA, B., BLANCH, H. W., WILKE, C. R.: Biotechnol. Bioeng. **25** (1983), 103.
[22] WORDEN, R. M., DONALDSON, T. L., SHUMATE, S. E.: Biotechnol. Bioeng. **18** (1983), 265.
[23] BLACK, G. M., WEBB, C., MATTHEWS, T. M., ATKENSON, B.: Biotechnol. Bioeng. **26** (1984), 134.

Book Review

A. BLAŽEJ, J. ZEMEK

Interbiotech '87
Enzyme Technologies
Progress in Biotechnology Vol. 4

Amsterdam, Oxford, New York, Tokyo: Elsevier Science Publishers. 1988, S. 535, $171.00
Dfl 325.00, ISBN 0-444-98924-2

The now available volume 4 of the series "Progress in Biotechnology" deals with the current state of enzyme engineering.
This volume contains review articles and original papers which have been presented at the international symposium INTERBIOTECH'87 in Bratislava. The editors have understood to connect the traditional lines of enzyme engineering with some new developments. They could present a good overview on this field which allows a quick insight not only for experts.
The following topics are included:

— Selection of enzyme producing microorganisms for different purposes
— Preservation of enzyme producing strains in culture collections
— Enzyme enrichment and purification methods including biospecific steps of purification
— Industrial application of immobilized biocatalysts
— Enzyme production with genetically modified microorganisms
— Modern aspects of enzyme engineering including protein engineering

On the basis of the presented material it is possible to get some stimuli for the own scientific work and for the valuation of the usefulness of enzymatic conversions in industrial scale.
The acquisition of the book is recommended for institutions with special interests in this field and for scientific libraries.

A. STEUDEL

Acta Biotechnol. 10 (1990) 1, 41—48 Akademie-Verlag Berlin

Isolation and Purification of Isoenzymes of Cellobiohydrolase I and II of Trichoderma reesei Using LPLC Methods

WITTE, K., HEITZ, H.-J., WARTENBERG, A.*

Fachrichtung Mikrobiologie, Universität des Saarlandes,
Universität — Bau 2, D-6600 Saarbrücken, F.R.G.

Summary

Five cellulases were fractionated from a commercial cellulase preparation (Celluclast™). Two isoenzymes of cellobiohydrolase I (CBHI) (pI = 4.1) could be proved to be real exo-glucanases due to their activity towards MU (= methylumbelliferyl)-lactoside being inhibited by cellobiose (5 mM) and due to production of cellobiose from carboxymethylcellulose (CMC) as the sole final product. Two isoenzymes of CBHII (pI = 6.15, 6.0) were shown to act as endo-glucanases because they produced glucose, cellobiose and cellotetraose from CMC and because they were not inhibited by cellobiose when decomposing MU-lactoside. Results confirm recent reports in the literature classifying CBHI and CBHII as exo-type and endo-type cellulases, respectively. Both the CBHI and the CBHII isoenzymes were shown to be active towards CMC and amorphous cellulose. CBHI and CBHII reactions could be differentiated from one another by the velocities of decomposition of CMC: CBHI acts slowly and linearly whereas CBHII acts strongly and exponentially. The fifth of the purified enzymes must be classed as a conventional endoglucanase which exhibits activity towards CMC but fails to be active towards MU-lactoside and amorphous cellulose.

Abbreviations: CBH, 1,4-β-D-glucan cellobiohydrolase (EC 3.2.1.91) or exo-glucanase; EG, 1,4(1,3;1,4)-β-D-glucan 4-glucanohydrolase or endo-glucanase; CMC, carboxymethylcellulose; MU-, 4-methylumbelliferyl-

Introduction

Fractionations of exo-type and endo-type cellulases produced by fungi present considerable problems. Progress in fractionating of these enzymes was made by using affinity chromatography [1, 2, 3, 4] or immunoaffinity chromatography [2, 5, 6]. This study shows that CBHI and CBHII of *Trichoderma reesei* can be fractionated into two isoenzymes in each case using common LPLC methods.

Materials and Methods

Enzyme Preparation

Celluclast™, a commercial cellulase preparation produced with *Trichoderma reesei* was kindly provided by NOVO INDUSRI AS (BAGSVAERD, Denmark). For chromatographic fractionation Celluclast proteins were purified by the following procedure. 500 ml of Celluclast were dilluted

* To whom correspondence should be addressed.

with distilled water to a volume of 5 l and dialyzed and reconcentrated by ultrafiltration (MINITAN-system, MILLIPORE) which was repeated four times. Proteins were then precipitated with 90 p.c. saturated $(NH_4)_2SO_4$ solution and washed with a saturated $(NH_4)_2SO_4$ solution. The precipitate was redissolved in water, desalted with a SEPHADEX G-25 column, reconcentrated with CX-10 MILLIPORE cartridges and finally lyophilized. For further purification by chromatographic fractionation the lyophilized material was redissolved to a concentration of 50 mg $\times$ ml^{-1} in 0.001 M KH_2PO_4 buffer, pH 7.4. Chromatographic methods are described in the Results and Discussions section. Isoelectric focusings were performed with precotes 3—10 (SERVA, Heidelberg) using a LKB Multiphor II.

Enzyme Assays

Activity towards CMC was measured by determining reducing sugars with the Nelson method [7] modified by Somogyi [8] using mixtures which contained 1 ml enzyme solution (1 mg $\times$ ml^{-1}) and 6 ml of a 0.25 p.c. Na-CMC solution (low viscosity, SIGMA) in acetate buffer (0.2 M, pH 4.8). The reaction of the experiment conducted at 40 °C was stopped by mixing 0.5 ml of the reaction mixture with 0.5 ml of Somogyi reagent. After tempering at 100 °C over 20 min. and recooling, 0.5 ml of Nelson reagent and 11 ml H_2O were added. Absorption was read at $\lambda = 660$ nm, which was then reduced by subtracting the absorption values of enzyme-free and of substrate-free control measurements. — Activity towards CMC was also determined viscosimetrically with a Ubbelohde viscosimeter (Schott) at 40 °C using reaction mixtures composed of 2.9 ml Na-CMC solution (0.25 p.c.) in 0.05 M acetate buffer (pH 4.8) and 0.1 ml enzyme solution (1 mg $\times$ ml^{-1}).
Activity towards amorphous cellulose was measured turbidimetrically using a Hach Ratio Turbidimeter 18900 (Test tube 18 mm). Values are expressed in Nephelometric Turbidity Units (NTU). The test tube part of the equipment was tempered at 40 °C with a Nicholson Air Stream Incubator. Measuring signals were transferred to an A/D-converter of an IBM XT 286 computer. Alltogether, 400 individual turbidity values (NTU) were taken within 12-second periods repeated 64 times during each experiment. Exponential functions were fitted to experimental data so that activity at zero time of incubation could be computed. — Amorphous cellulose was prepared with the following procedure. 2.5 g cellulose microcrystalline 0.018 mm SERVA (Heidelberg, FRG) were completely dissolved in H_3PO_4 (85 p.c.) at 20 °C. To reprecipitate cellulose the solution was subsequently dropped into 400 ml of distilled water vigorously stirred by a Waring blender. The precipitate was filtered by using a suction filter, resuspended and filtered out again until the pH of the filtrate became neutral. After storing the resuspended cellulose over night in a refrigerator, the low density supernatant fraction of reprecipitate was subjected to desintegration by using a *Sonifier* B-12 (BRANSON Sonic Power Company; 25 treatments at 75 W executed in an ice-bath using the macro tip and lasting 15 sec. interrupted by periods of 25 sec. each). Right before an experiment, cellulose solution was filtered once more and diluted to the desired density with distilled water.
Activities towards MU-glycosides (MU-glucopyranoside, MU-cellobioside) were determined photometrically at $\lambda = 365$ nm. Reaction mixtures containing 2 ml of a 0.3 mM solution of one of the MU-glycosides in acetate buffer (0.05 M, pH 4.8) and 50 µl enzyme solution (1 mg $\times$ ml^{-1}) were tempered at 40 °C over 30 min. The reaction was stopped by mixing 0.5 ml of reaction mixture with 3.5 ml glycine buffer (0.5 M, pH 10.4). Activity values are expressed in International Units (U; [µg $\times$ mg^{-1} $\times$ min^{-1}]). Activity towards MU-lactoside was determined qualitatively for testing fractions of all peaks manifested during the chromatographic purification steps (Fig. 1) using the same composition of reaction mixture with and without an addition of 5 mM cellobiose.

Thinlayer Chromatography of Products Yielded from CMC Decomposition

The incubation mixture was composed of 500 µl of CMC solution (2 p.c. CMC, SIGMA, medium viscosity, in 0.05 M Na-acetate buffer, pH 4.8) and 25 µl of enzyme solution (0.1 p.c.). Control experiments were performed at pH 4.0 and 5.5. Incubations were executed at 40 °C. Samples were taken after 1, 8, 24 and 48 h. HPTLC plates (Si60 F_{254}, 10 $\times$ 20 cm), provided with a concentration zone, were used. Application: 1 µl of sample solution in each case. Solutions of reference sugars contained 3 p.c. of Cellodextrine Mixture (MERCK) or 2 p.c. glucose or cellobiose. Development: butanol/methanol/water (5:5:3).

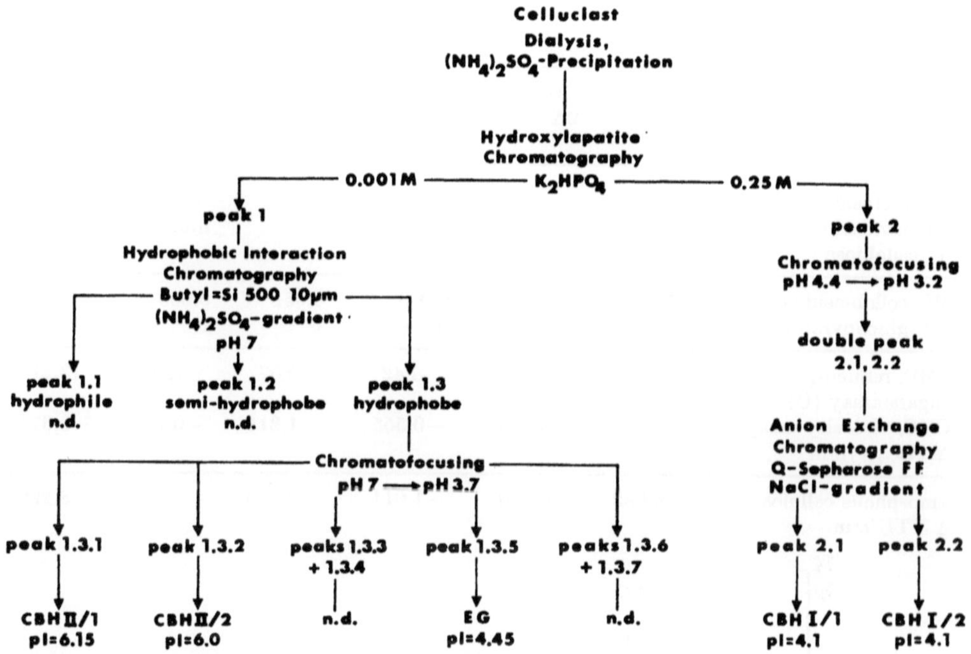

Fig. 1. Flow sheet of the chromatographic purification steps of Celluclast cellulases

Results and Discussion

The procedure conducted for fractionating partially purified Celluclast is reflected by Fig. 1. The strategy of the fractionating procedure was to find homogenous protein fractions which are marked out by pI-values in the ranges of 3.8 to 4.3 and of 5.9 to 6.3 being characteristic for *Trichoderma reesei* CBHI and CBHII, respectively [3, 5, 9—12]. After each step isoelectric focusings were performed for testing protein composition, homogeneity and pI-values of all resulting peaks. Additionally, diverse fractions of each peak were tested for activity towards MU-lactoside, with and without cellobiose as an inhibitor, for differentiating betwen CBHI and CBHII. CBHI, as a real exo-glucanase, releases lactose from MU-lactoside whose reaction is strongly inhibited by cellobiose [11] whereas CBHII acting as an endo-glucanase [9] was reported to fail in decomposing MU-lactoside [11]. Contrary to this latter result CBHII was shown to be active towards MU-lactoside in our experiments but the reaction failed to be inhibited by cellobiose, being in conformance with the result that inhibition of MU-lactoside decomposition as induced by cellobiose allows a distinction between CBHI and EG [11]. Results of essential MU-lactoside assays are given in Tab. 1.

Hydroxylapatite chromatography was started with 0.001 M KH_2PO_4 buffer (pH 7.4) yielding peak 1 as a direct flow out from the column. Increase of buffer concentration to 0.25 M was followed by a flow out of peak 2. Peak 1 after reconcentration was further treated by hydrophobic interaction chromatography which produced peak 1.1 to 1.3 from which only the latter one showed activity towards MU-lactoside. Peak 1.3 subjected to chromatofocusing (pH 7 to 3.7) was fractionated into seven peaks (1.3.1 to 1.3.7, Fig. 2) from which the first and the second one revealed activities towards MU-lactoside without being inhibited by cellobiose (Tab. 1, CBHII/1, CBHII/2). Isoelectric focusing

Tab. 1. Activities of the cellulases yielded homogenously from Celluclast by chromatographic purification

enzyme (peak) → substrate ↓	CBHI/1 2.1	CBHI/2 2.2	CBHII/1 1.3.1	CBHII/2 1.3.2	EG 1.3.5	Celluclast
1 MU-lactoside	+	+	+	+	−	+
2 MU-lactoside + cellobiose	−	−	+	+	n.d.	n.d.
3 MU-cellobioside (U)	0.063	0.074	0.008	0.051	0.006	0.117
4 MU-glucopyranoside (U)	0	0	0	0	0	0.037
5 CMC, reducing sugars assay (U)	3.35	3.35	8.42	10.07	7.00	13.21
6 CMC, viscosity assay Δ visc.$_{rel.}$/min	−0.019	−0.013	−0.555	−1.211	−0.119	−0.673
7 amorphous cellulose Δ NTU/min	−0.582	−0.730	−1.011	−2.976	−0.096	−8.299

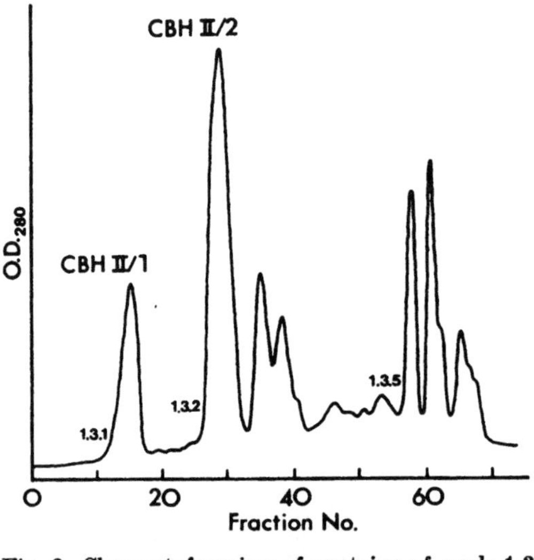

Fig. 2. Chromatofocusing of proteins of peak 1.3 yielded from hydrophobic interaction chromatography (cf. Fig. 1) resulting in homogenous peaks which contain CBHII/1, CBHII/2 and an endo-glucanase

(Fig. 4) proved homogeneity of the two peaks and in addition of peak 1.3.5 (pI = 6.15, 6.0 and 4.45, respectively). Peak 1.3.5 failed to be active on MU-lactoside. For the present proteins of peak 1.3.1 and 1.3.2 can be claimed to belong to CBHII, whereas the protein of peak 1.3.5 should be considered to be an endo-glucanase, which fails to be active towards MU-lactoside.

Peak 2 from the hydroxylapatite chromatography showed activity towards MU-lactoside which decreased to zero when inhibited by cellobiose, thus giving reason to deduct that peak 2 proteins belongs to CBHI [11]. Chromatofocusing of peak 2 resulted in a double peak which still contained an impurity lying between the two CBHI component

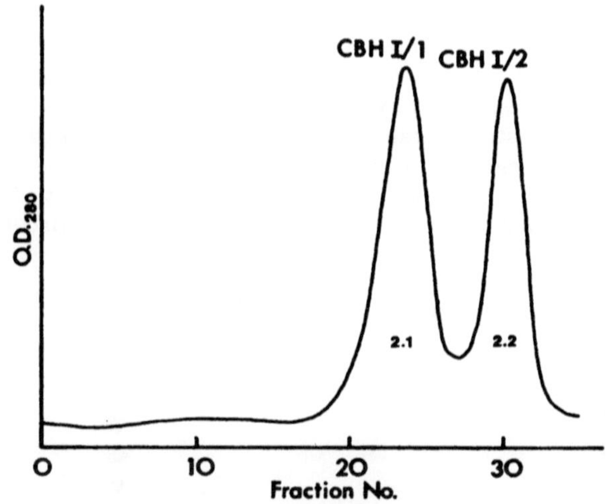

Fig. 3. Anion exchange chromatography of proteins of the double peak 2.1/2.2 resulting in homogenous peaks which contain CBHI/1 and CBHI/2

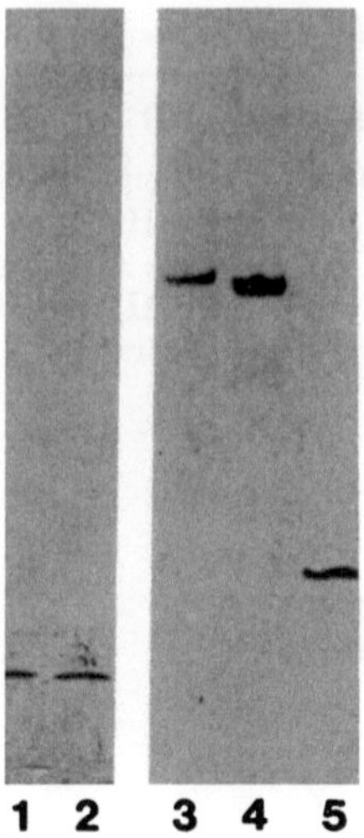

Fig. 4. Analytical isoelectric focusings of the proteins which resulted homogenously from chromatographies outlined in Fig. 1.
Lanes 1 and 2 = CBHI/1 and CBHI/2 resp., lanes 3 and 4 = CBHII/1 and CBHII/2, resp., lane 5 = E 6

peaks. The impurity could be separated from the CBHI components and the components could be separated from one another by submitting the double peak material to anion exchange chromatography (Fig. 3). Materials from the resulting peaks (2.1, 2.2) were proved to be active towards MU-lactoside and became inactive with regard to MU-lactoside decomposition when cellobiose was added to the reaction mixture (Tab. 1). Isoelectric focusing revealed homogeneity of the two proteins and identical pI-values (pI = 4.1; Fig. 4) so that the proteins from peak 2.1 and 2.2 can be classified as CBHI isoenzymes.

The classifications of the enzymes of peak 2.1 and 2.2 (Fig. 3) as CBHI isoenzymes and those of peak 1.3.1 and 1.3.2 (Fig. 2) as CBHII isoenzymes were verified by determining the products of CMC decomposition (Fig. 5). Cellobiose turned out to be the sole end-

product when CMC was decomposed by fraction 2.1 and 2.2 whereas in the case of CMC decompositions induced by peak 1.3.1 and 1.3.2 as well as 1.3.5, glucose, cellobiose and cellotetraose were produced indicating a random type of cellulose decomposition which is characteristic for endo-glucanase activity.
To confirm deductions hitherto stated, fractions of the five peaks which turned out to be homogenous with regard to isoelectric focusing were tested for activities towards CMC and amorphous cellulose. Peak 1.3.1, 1.3.2, 2.1 and 2.2 exhibited activity towards both substrates, CMC and amorphous cellulose (Tab. 1). Fig. 6 shows that decomposition of

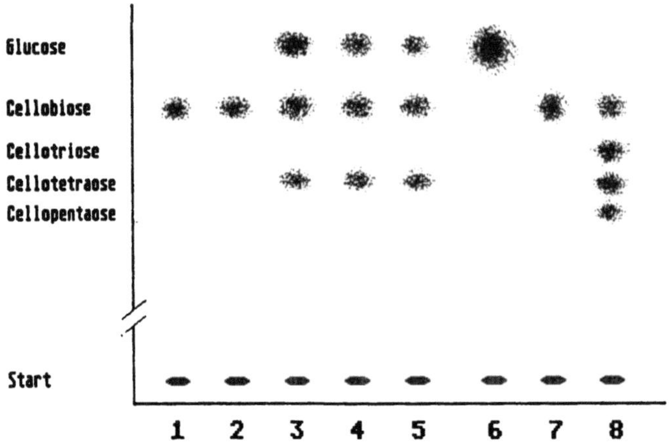

Fig. 5. Computer graphic reproduction of a thinlayer chromatography of reaction products resulting from decomposing CMC by cellulases (pH 4.8) as purified from Celluclast. Experiments executed at pH 4.0 and 5.5 gave the same products as the pH 4.8 experiment outlined in this figure. Lanes 1 to 5 = CBHI/1, CBHI/2, CBHII/1, CBHII/2, EG from peak 1.3.5; lanes 6 to 8 = reference sugars

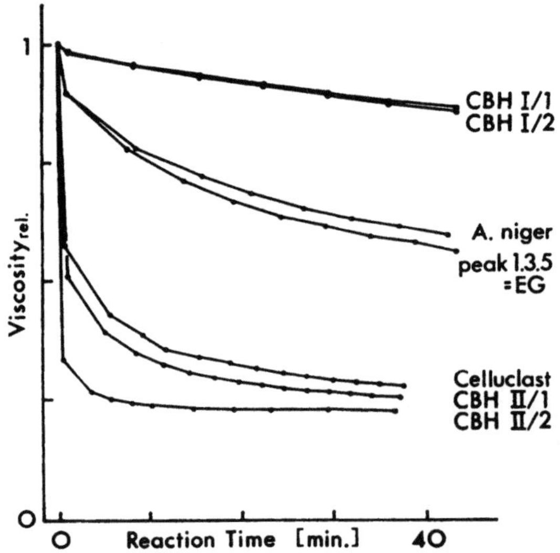

Fig. 6. Decompositions of CMC catalyzed by the cellulases (62.5 μg/ml protein) purified from Celluclast. Control measurements were performed with non-fractionated purified Celluclast (62.5 μg/ml) and with *Aspergillus niger* cellulase (125 μg/ml)

CMC catalyzed by peak 2.1 and 2.2 run slowly and nearly linear when plotted versus time whereas decompositions driven by peak 1.3.1 and 1.3.2 proteins run very exponentially. This confirms the deduction given above: The proteins of peak 2.1 and 2.2 must be isoenzymes of CBHI. CBHI is known to act really as an exo-glucanase, releasing only cellobiose from amorphous cellulose [2] so that attack of this enzyme to cellulose should be followed by a linear shortening of cellulose chains. In contrast, decompositions of cellulose driven by endo-glucanases should be expected to run exponentially. The result that CBHII/1 and CBHII/2 (peak 1.3.1, 1.3.2) decompose CMC whose reactions are related to those of exponential kinetics corresponds to the finding that CBHII acts as an endo-glucanase named EGIb [9]. The difference between the two groups of cellulases (CBHI and CBHII) is additionally reflected by the values of reducing sugars released from CMC (Tab. 1): Activities of CBHII isoenzymes were three times higher compared with the activities of CBHI isoenzymes.

The enzyme given in peak 1.3.5 was shown to be active towards CMC (release of reducing sugars, viscosity assay) and MU-cellobioside (Tab. 1). No activity was found against amorphous cellulose and MU-lactoside (Tab. 1). It acts towards CMC (viscosity assay, Fig. 6) with a kinetics corresponding with that of *Aspergillus niger* cellulase (Fig. 6) which is unable to attack amorphous cellulose noticeably [10]. This enzyme must therefore be classed as one of the conventional endo-glucanases whose deduction is confirmed by the pI-value of the enzyme (4.45) lying in the range (pH 4 to 6.3) which is known to be characteristic for most of the endo-glucanases of *Trichoderma reesei* [11, 13]. Furthermore, the deduction is confirmed by detecting glucose, cellobiose and cellotetraose as reaction products when CMC was attacked by peak 1.3.5 enzyme (Fig. 5). — For completion, no one of the five studied enzymes showed activity towards MU-glucopyranoside (Tab. 1) giving evidence that the preparations were free from β-glucosidase activity.

We mean that activities of exo-cellulases and endo-cellulases cannot further be related alternatively with decompositions of amorphous cellulose and CMC, respectively. Each of the two substrates can be decomposed by exo-glucanases and endo-glucanases. The difference between exo- and endo-activity is not caused by the specification for one of the two substrates but is expressed by the velocities of decompositions being obviously controlled by the different reactivities of the two types of cellulases. Activity of endo-glucanases attacking CMC but not attacking amorphous cellulose is really only characteristic of conventional endo-glucanases but not of those represented by CBHII/1 and CBHII/2. The kinetics of these non-conventional endo-glucanases regarding decomposition of amorphous cellulose will be published in another report.

Received January 23, 1989

Acknowledgements

The authors are grateful to Mr. P. MEIERS for his assistance in carrying out some experiments and editing computer programs and to Mr. J. POCKLINTON for proofreading the manuscript.

References

[1] NUMMI, M., NIKU-PAAVOLA, M.-L., ENARI, T.-M., RAUNIO, V.: Anal. Biochem. **116** (1981), 137.

[2] NUMMI, M., NIKU-PAAVOLA, M.-L., LAPPALAINEN, A., ENARI, T.-M., Raunio, V.: Biochem. J. **215** (1983), 677.

[3] VAN TILBEURGH, H., BHIKHABHAI, R., PETTERSSON, L. G., CLAEYSSENS, M.: FEBS Lett. 169 (1984), 215.
[4] WEBER, M., FOGLIETTI, M. J., PERCHERON, F.: J. Chromatogr. 188 (1980) 377.
[5] NIKU-PAAVOLA, M.-L., LAPPALAINEN, A., ENARI, T.-M., NUMMI, M.: Biotechnol. Appl. Biochem. 8 (1986), 449.
[6] RISKE, F. J., EVELEIGH, D. E., MACMILLAN, J. D.: J. Ind. Microbiol. 1 (1986), 259.
[7] NELSON, N.: J. Biol. Chem. 153 (1944), 376.
[8] SOMOGYI, M.: J. Biol. Chem. 195 (1952), 19.
[9] KYRIACOU, A., MACKENZIE, R. C., NEUFELD, J. R.: Enzyme Microb. Technol. 9 (1987), 25.
[10] HEITZ, H. J., WITTE, K., WARTENBERG, A.: Acta Biotechnol. 10 (1990) 3 (in press).
[11] VAN TILBEURGH, H., CLAEYSSENS, M.: FEBS Lett. 187 (1985), 283.
[12] VAN TILBEURGH, H., PETTERSSON, G., BHIKABHAI, R.: DE BOECK, H., CLAEYSSENS, M.: Eur. J. Biochem. 148 (1985), 329.
[13] SHEIR-NEISS, G.; MONTENECOURT, B. S.: Appl. Microbiol. Biotechnol. 20 (1984), 46.

Book Review

H. M. KALISZ

Microbial Proteinases

M. MAESTRACCI, K. BUI, A. THIERY, A. ARNAUD, P. GALZY

The Amidase from a Brevibacterium Strain: Study and Applications

aus der Schriftenreihe A. FIECHTER

Advances in Biochemical Engineering/Biotechnology

Vol. 36 **Enzyme Studies**

Berlin: Akademie-Verlag, 1988. 115 S., M 108,—, ISBN 3-05-500568-6

Der vorliegende Band 36 der obengenannten Schriftenreihe setzt ihr Anliegen fort, zur Verbreitung der Biotechnologie bzw. biotechnologischer Methoden einen Beitrag zu leisten.
Dieser Band enthält Übersichtsbeiträge über zwei biotechnologisch bedeutsame Enzymgruppen, und zwar über mikrobielle Proteinasen sowie über mikrobielle Amidasen. Die publizierten Einzelbeiträge sind gut gegliedert und anschaulich gestaltet. Dadurch ist es möglich, sich rasch einen Überblick über den gegenwärtigen Stand der Forschungen auf diesem Gebiet zu verschaffen. Die zu den Beiträgen gehörigen Literaturübersichten mit hohem Aktualitätsgrad erleichtern dem interessierten Leser ein schnelleres Eindringen in die Gesamtproblematik. Dem einschlägig orientierten Praktiker gibt dieser Band eine gute Basis für weiterführende Überlegungen zur Erschließung neuer Anwendungsfelder für diese Enzyme im eigenen Praxisumfeld.
Darüber hinaus kann aber auch der Fachmann die sehr aktuellen Übersichtsbeiträge als Nachschlagemöglichkeit nutzen bzw. daraus Anregungen für eigene zukünftige Arbeiten entnehmen.
Alles in allem stellt der Band 36 eine gute Erweiterung dieser Schriftenreihe dar und ist allen biotechnologisch Interessierten in Forschung und Praxis zu empfehlen. Auch für die Einbeziehung in die Studentenausbildung der Fachrichtungen Biochemie/Biotechnologie ist das vorliegende Buch geeignet.

A. STEUDEL

Acta Biotechnol. 10 (1990) 1, 49–53 Akademie-Verlag Berlin

Fructose Phosphorylation by Zymomonas mobilis Glucokinase

EDYE, L. A., KOSITANONT, CH., DOELLE, H. W.*

Department of Microbiology, University of Queensland,
St. Lucia-Brisbane, Qld. 4067, Australia

Summary

Zymomonas mobilis ATCC 53431, a fructokinase negative mutant is unable to utilize fructose as a sole carbon source for growth. At high fructose concentrations, however, fructose was converted to ethanol. The fructose uptake displayed MICHAELIS-MENTEN kinetics with an apparent *Km* of 185 mM fructose.
Purified glucokinase from ATCC 53431 and the wild strain ATCC 29191 both exhibited fructose phosphorylating activity at high fructose concentrations with an apparent *Km* value of 222 mM fructose. Glucokinase substrate specificity was found not to be absolute, as previously reported.

Introduction

The carbohydrate metabolism of *Zymomonas mobilis* has been reviewed [1]. The recent interest in this bacterium is concomitant with its realized industrial potential for ethanol production.
Zymomonas mobilis transports glucose by a facilitated diffusion system [2, 3], and phosphorylates glucose and fructose by separate specific kinases, viz. glucokinase and fructokinase [4–6]. Studies on these kinases conclude that both enzymes are absolutely specific for their respective carbohydrate substrates.
Several publications report the use of *Zymomonas mobilis* strains to accumulate fructose in fermentations of sucrose or glucose/fructose mixtures with the aim of producing both ethanol and a commercial fructose sweetener [7–9]. BRINGER-MEYER et al. [7] used *Zymomonas mobilis* mutants blocked in fructose utilization to ferment glucose/fructose mixtures. The mutant strain was shown to lack fructokinase activity. During investigations into the fermentation of sucrose using the fructokinase negative mutant *Z. mobilis* ATCC 53431, a possible 'leakage' of fructose to ethanol was observed at high sucrose concentrations (350 g/l). Excessive ethanol yields were obtained with below theoretical fructose yields despite the fact that no mutant reversion or contamination occurred.
This paper reports on the nature of this fructose loss during sucrose fermentation, viz. fructose phosphorylation by glucokinase at high fructose concentrations.

* To whom correspondence should be addressed.

Materials and Methods

Culture Maintenance. The fructose utilization negative mutant *Zymomonas mobilis* ATCC 53431 derived from *Z. mobilis* ATCC 39676 (US Patent 4797360) [9] was maintained at 4 °C in a maintenance medium containing 50 g/l glucose and 2 g/l each of $(NH_4)_2SO_4$, KH_2PO_4, $MgSO_4 \times 7\,H_2O$, yeast extract and casein hydrolysate. The cultures were kept on agar plates and liquid cultures and subcultured weekly. Metabolic reversion was monitored by patch testing, carried out once per week using glucose and fructose agar plates containing 10 g/l of the respective sugar in the maintenance medium.

Fructose Uptake. Maintenance medium containing either 100 g/l sucrose or 100 g/l glucose were inoculated with 10% (v/v) preculture and incubated statically at 32 °C. After 12 hrs incubation, six equal portions (6 × 200 ml) of each culture were centrifuged for 15 min at 2,000 × *g*, washed in saline and re-centrifuged. The cell pellets were resuspended in maintenance medium containing increasing fructose concentrations (10–200 g/l). The liquid cultures were again incubated statically at 32 °C. Samples were taken initially and after 4 hrs incubation for fructose and ethanol determination according to DOELLE & GREENFIELD [8]. All final samples were patch tested for reversion and contamination.

Enzyme Purification. After 15 hrs incubation at 37 °C, cells grown in the maintenance medium containing 10% sucrose, were harvested for 15 min at 4,000 rpm using a refrigerated International B-20 centrifuge. The resulting pellets were washed twice with a buffer containing 50 mM Tris-HCl (pH 7.0), 10 mM $MgCl_2$, 0.1 mM EDTA and 2 mM mercaptoethanol. After washing, the cells were resuspended in 30 ml of the Tris-HCl buffer and cell free extracts were obtained as described by DOELLE [5].

Separation, purification and enzyme activity determinations were carried out as described by DOELLE (1982a, b), with the exception that DEAE-Trisacryl M was used instead of DEAE-cellulose.

The eluted enzyme fractions were concentrated by ultrafiltration using Amicon PM-10 ultrafilter membranes. Concentrated enzyme solution was applied to a Sephacryl S-200 column (1.6 × 50 cm). The highest enzyme activity fractions were pooled and used for all kinetic studies and electrophoresis.

Non-denaturing discontinuous electrophoresis was carried out at neutral pH on 0.5 mm polyacrylamide gels using buffers described by HAMES [10]. Proteins were stained using the silver staining technique [11], whereas the glucokinase activity stain used Nitroblue Tetrazolium and Phenazine metho sulfate linked through glucose 6-phosphate dehydrogenase [12].

Enzyme Assays. Glucokinase and fructokinase activity determinations were carried out according to DOELLE [5]. Fructose phosphorylation assays contained 2 M fructose in 50 mM Tris-HCl buffer (pH 7.0) using the fructokinase and the glucokinase assay methods. The reaction was initiated by the addition of 0.05 ml of suitably diluted purified glucokinase extract.

For comparative purposes a purified glucokinase preparation was obtained from Dr. R. K. SCOPES, La Trobe University, Bundoora, Australia, originating from *Zymomonas mobilis* ATCC 29191, a wild type strain [6].

Results

Fructose Uptake Using Whole Cells. Zymomonas mobilis ATCC 53431 cells precultured in maintenance medium containing sucrose were able to slowly metabolize fructose to ethanol (Tab. 1) when suspended in fructose solutions. In all cases, these cultures were unable to utilize fructose for growth on agar plates. The reciprocal plot of fructose concentration versus fructose uptake rate by the sucrose precultured cells (Fig. 1) fits MICHAELIS-MENTEN kinetics with an apparent *Km* for fructose of 185 mM and a substrate inhibition above 125 g/l fructose.

Cells precultured on maintenance medium containing glucose, recorded much lower fructose uptake rates (Tab. 1) and ethanol was not detectable in the medium.

Tab. 1. The effect of fructose concentration on fructose uptake using whole cells of *Zymomonas mobilis* ATCC 53143 precultured in sucrose or glucose containing maintenance media

Fructose	Sucrose Preculture		Glucose Preculture	
	$-d$ fruc d time	ethanol produced	$-d$ fruc d time	ethanol produced
[g/l]	[g/l/h]	g/l	[g/l/h]	g/l
10.00	0.86	1.30	0.04	0.0
50.00	1.75	3.36	0.19	0.0
100.00	2.15	4.38	0.24	0.0
125.00	2.34	4.59	0.20	0.0
200.00	0.90	1.80	0.21	0.0

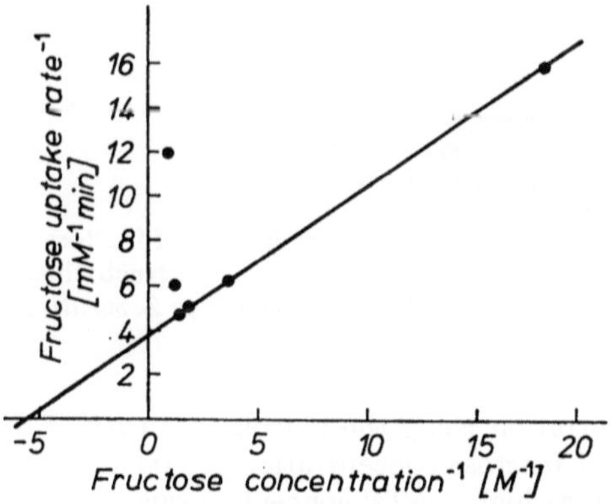

Fig. 1. Reciprocal plot of fructose uptake rates using whole cells of the fructokinase negative mutant *Zymomonas mobilis* ATCC 53143 precultured on sucrose containing maintenance medium

Fructose Phosphorylation by Purified Glucokinase Glucokinase extracts from two strains of *Zymomonas mobilis* were investigated, strain ATCC 53431, a fructokinase negative mutant, and a wild type strain ATCC 29191. The fructose phosphorylating activities of both purified enzyme extracts exhibited linear relationships to protein concentrations up to 8 or 10 μg protein and exhibited a single protein and activity band after gel electrophoresis.

Comparative rates of glucose and fructose phosphorylation using the purified enzyme extracts are shown in Tab. 2. Fructose phosphorylation was not measurable at levels

Tab. 2. Comparative phosphorylation rates of glucose and fructose using *Zymomonas mobilis* glucokinase extracts

Substrate	Phosphorylation Rates [mM/min/mg]	
	ATCC 53143	ATCC 29191
5 mM Glucose	3.44	4.39
5 mM Fructose	< 0.05	< 0.05
800 mM Fructose	1.12	1.38

of 5 mM fructose, although significant rates were obtained at 800 mM fructose. Fructose phosphorylation rates became measurable at fructose concentrations above 100 mM. Reciprocal plots of fructose concentration versus glucokinase velocity (Fig. 2) fit MICHAELIS-MENTEN kinetics with an apparent *Km* value of 222 mM fructose for ATCC 2919 and ATCC 53431.

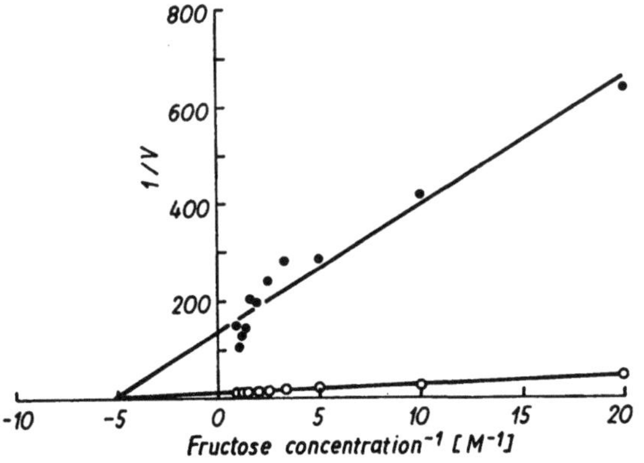

Fig. 2. Reciprocal plot and apparent *Km* determination of fructose phosphorylation by purified glucokinase extracts (3 µg protein) from the fructokinase negative mutant *Z. mobilis* ATCC 53143 (•; 1/v × 10) and the wild type *Z. mobilis* ATCC 29191 (○)

The reciprocal plot of ATP concentration using the glucokinase extract of *Zymomonas mobilis* ATCC 29191 at the constant fructose concentration of 100 mM exhibit an apparent *Km* of 0.44 mM and inhibition at higher ATP concentrations.
A Mg: ATP ratio of 10:1 had no significant effect on the activity.

Discussion

According to ROMANO et al. [2], *Zymomonas mobilis* has no PEP-glucose phosphotransferase system and no permease system. Instead a constitutive stereospecific low affinity high velocity facilitated diffusion system for glucose transport has been reported [3, 13]. The monomeric sugars carried into the cell are then phosphorylated by constitutive and substrate specific kinases [4—6]. At 10 mM concentrations of carbohydrates, glucokinase was unable to phosphorylate fructose and likewise, fructokinase was unable to phosphorylate glucose [6]. Further, the effects of glucose on fructokinase phosphorylation of fructose and fructose on glucokinase phosphorylation of glucose were tested up to 111 mM concentrations (4—6). No phosphorylation of fructose by glucokinase or of glucose by fructokinase was detected.
The presented results show that fructose can be phosphorylated by glucokinase at higher than 100 mM fructose concentrations. Both glucokinases, from the wild type and from the fructokinase negative mutant, exhibit fructose phosphorylation under the conditions stated. It is notable that the apparent *Km*'s for fructose phosphorylation by glucokinase calculated using the two pure enzymes is in agreement with the apparent *Km* calculated in the whole cell experiment. The very slow phosphorylation of fructose by glucokinase must determine the rate of fructose 'leakage' in fructokinase negative *Zymomonas mobilis* fermentations of sucrose.

Significant fructose uptake by the fructokinase-negative mutant only occurs when the organism is precultured in medium containing sucrose. It is tempting to speculate that the effect of carbohydrate source in preculturing on fructose uptake is evidence for fructose transport via carrier-mediated diffusion. Such transport would then be independent from the organism's potential to further utilize this sugar via phosphorylation. This facilitated diffusion transport system must be inducible by sucrose. It would appear that a fructose transport system can be induced without a fructokinase present, furthermore it may be possible that glucose-grown cells possess the fructokinase without the fructose transport system. This would explain the retarded fructose metabolism in fructose containing medium of glucose-grown precultures, compared to sucrose or fructose-grown precultures.

Acknowledgements

We greatly appreciate the financial help received from the University of Queensland Research Funds together with the scholarship obtained from IDP (C.K.). We are indebted to the generosity of Dr. R. K. SCOPES, LaTrobe University, Bundoora, for sending us his purified glucokinase from *Zymomonas mobilis* ATCC 29191.

Received January 31, 1989

References

[1] VIIKARI, L.: CRC Crit. Rev. Biotechnol. **7** (1988), 237.
[2] ROMANO, A. H., TRIFONE, J. D., BRUSTOLON, M.: J. Bacteriol. **139** (1979), 93.
[3] DIMARCO, A. A., ROMANO, A. H.: Appl. Environ. Microbiol. **49** (1985), 151.
[4] DOELLE, H. W.: Eur. Appl. Microbiol. Biotechnol. **15** (1982), 20.
[5] DOELLE, H. W.: Eur. J. Appl. Microbiol. Biotechnol. **14** (1982), 241.
[6] SCOPES, R. K., TESTOLIN, V., STOTER, A.: Biochem. J. **228** (1985), 627.
[7] BRINGER-MEYER, S., SCOLLAR, M., SAHM, H.: Appl. Microbiol. Biotechnol. **23** (1985), 134.
[8] DOELLE, H. W., GREENFIELD, P. F.: Appl. Microbiol. Biotechnol. **22** (1985), 411.
[9] SUNTINANALERT, P., PEMPERTON, J. P., DOELLE, H. W.: Biotechnol. Lett. **8** (1986), 351.
[10] HAMES, B. D. — In: Gel Electrophoresis of Protein: a Practical Approach (HAMES, B. D., DICKWOOD, D., Eds.), 1981, IRL Press Ltd. Oxford.
[11] BLUM, H., BEIER, H., GROSS, H. J.: Electrophoresis **8** (1987), 93.
[12] EWINGS, K. N. — In: On the isoenzymes of phosphofructokinase from some Enterobacteriaceae. Ph. D. thesis, 1980, University of Queensland, Brisbane.
[13] BELAICH, P., SENEZ, J. C., MURGIER, M.: J. Bacteriol. **95** (1968), 1750.

Book Review

LEVY, S. B.; R. P. NOVICK (Eds.)

Antibiotic Resistance Genes: Ecology, Transfer, and Expression

Banbury Report 24
Cold Spring Harbor: Cold Spring Harbor Laboratory, 1987, $ 68,00, XIV, 436 pp.
ISBN 0-87969-224-3

Im Frühjahr 1986 trafen sich im Banbury Center des Cold Spring Harbor Laboratory, unterstützt durch die „U.S. Environmental Protection Agency" und die „National Science Foundation", Wissenschaftler molekularbiologischer, biologischer, genetischer und ökologischer Wissensgebiete zum Erfahrungsaustausch über Fragen, die den möglichen Ursprung von Antibiotika-Resistenzgenen und ihre natürliche Verbreitung in der Umwelt betreffen. Es sollten allgemeine Vorstellungen zur Genverteilung und Evolution erarbeitet werden. Die dazu gesammelten Daten werden bei der Analyse der möglichen Verteilung anderer Gene in anderen Organismen im Labor oder in der natürlichen Umwelt helfen. Solche Informationen können auch bei der Abschätzung des Verhaltens von genetisch modifizierten Organismen oder genetischen Elementen in der Natur hilfreich sein.

Beginnend mit der Definition der Species, die am Austausch von Antibiotika-Resistenzgenen teilnehmen, den Beiträgen zu genetischen Elementen, die diese Gene tragen bzw. für deren Mobilität verantwortlich sind, wenden die Autoren sich schließlich diesen Genen selbst zu. Dabei werden Themen gestreift, die den Ursprung und die Evolution dieser Gene, ihre Transportvesikel und Faktoren, die ihre Verteilung und Evolution bedingen, beleuchten.

Es ist erstaunlich, daß trotz der Vielfalt der dargestellten Probleme und der großen Themenpalette dieses Buch einen äußerst geschlossenen Eindruck hinterläßt. Ebenfalls positiv ist hervorzuheben, daß jedem Beitrag die Kommentare der Tagungsteilnehmer angefügt sind. Die durchweg fundierten und recht aktuellen Artikel sind gut und übersichtlich bebildert und mit hilfreichen Literaturverzeichnissen versehen. Den Schluß bilden ein Namens-Index und ein umfangreiches Stichwortverzeichnis.

Es ist zu vermuten und zu hoffen, daß dieses Buch sich einen breiten Leserkreis, besonders unter den biologisch und medizinisch orientierten Naturwissenschaftlern erobern wird.

E. SPANIER

Acta Biotechnol. 10 (1990) 1, 55–61 Akademie-Verlag Berlin

Ethanolic Fermentation with Saccharomyces cerevisiae Cells Immobilized in Pectate Gel

RÜHLEMANN, I., RICHTER, K., BERGER, R.

Academy of Sciences of the G.D.R.
Institute of Biotechnology
Permoserstraße 15, Leipzig, 7050 G.D.R.

Summary

High-molecular weight pectic acid with a STAUDINGER index of 210 ml/g and a degree of esterification of 3% was used as matrix material for the immobilization of *Saccharomyces cerevisiae* cells. In discontinuous and continuous fermentation tests the gel beads obtained exhibited the same biomass loading capacity (152–155 g dry wt. cells/kg gel) and about the same maximum specific productivity (103.0 g ethanol/kg gel · h) as alginate immobilizates. But there were distinct differences in the swelling behaviour of the two gels. Under the same experimental conditions the increase of bead volume amounted to 27% only for pectate gel in comparison to 129% for alginate gel. In continuous fermentation experiments performed in a horizontal-column packed-bed reactor with liquid recycling a mean steady-state ethanol concentration of 69.1 g/l and a mean productivity of 24.7 g ethanol/lh could be kept constant over a period of more than 10 days.

Introduction

The fundamental suitability of pectinate gels as immobilization matrices was discussed by VORLOP [1] previously. He compared some special properties of Ca and Mg pectinate gels with those of Ca alginate and found that the pectinate beads exhibited a higher ion tolerance and a much lower swelling capacity compared to Ca alginate beads. The mechanical stability was like that of alginate beads, provided that the biomass loading of the gel did not exceed a value of 20% (wet mass).
The application of pectinate-entrapped yeast cells in alcoholic fermentation has only been reported by NAVARRO et al. [2, 3] and NOGUCHI et al. [4–6] till now. NAVARRO [3] performed continuous fermentation experiments in a wedge-shaped horizontal reactor with plug-flow characteristics using a citrus pectine as immobilization matrix and molasses as substrate. Under steady-state conditions a mean ethanol concentration of 64 g/l and a mean productivity of 17 g/lh referred to the working volume of the reactor were obtained. Probably caused by both diffusion resistances and extra inhibitory effects a loss of process stability was observed after 11–12 days steady-state fermentation and the fermentative activity of the immobilized cells depended significantly on its location in the reactor. But detailed data about the characteristics of the pectinate gels used were not given by the authors mentioned above.

On the contrary we could show in previous papers [7, 8] that Ca or Al pectate beads prepared by use of high-molecular weight pectic acid (STAUDINGER index > 200 ml/g) having a low methoxy group content (< 3%) are well suited for the immobilization of yeasts and bacteria. The pectate gels showed a much higher tolerance against both complexing agents (phosphate, citrate, lactate, tartrate etc.) and other gel-destabilizing ions (Na^+, K^+, NH_4^+ etc.) than alginate beads. Therefore pectate-entrapped microorganisms can be used advantageously in special fermentations involving high concentrations of those compounds in the culture medium. Furthermore one may suppose a similiar performance behaviour of beads formed from pectate and alginate in processes with media poor in destabilizers.
In this paper some results are presented which were obtained from comparative ethanolic batch fermentation experiments carried out with pectate and alginate gels as immobilization matrices. Moreover, the performance of a continuous pectate-immobilized cell horizontal column reactor with liquid recycling is reported.

Materials and Methods

Microorganism

The yeast *Saccharomyces cerevisiae* IBT H 191 from the culture collection of our institute was used in this research.

Immobilization

The immobilization of the yeast cells was performed in the way described previously [8] using a pectic acid with a STAUDINGER index of 210 ml/g and a degree of esterification of 3% and Na alginate from RIEDEL DE HAËN (STAUDINGER index: 711 ml/g). The immobilization mixture contained 3 g pectic acid (or 1 g Na alginate) and 10.5 g wet biomass per 100 ml. As precipitating bath 0.1 M $CaCl_2$ was used unless the beads were provided for long-time experiments, in which case 0.4 M $CaCl_2$ was applied. The diameter of the beads was varied between 1.1 and 3.2 mm.

Apparatus and Process Parameters

The discontinuous experiments and some continuous short-time fermentations were carried out in a 0.4 l glass reactor equipped with a magnetic stirrer and devices for temperature and pH-value control. Both process parameters were kept constant at T = 32 °C and pH = 4.2. The volume fraction of gel beads in the working volume (0.25 l) was varied within the limits of 0.2 to 0.34.
In each batch process the liquid was removed from the fermenter after finishing the fermentation and the remaining beads were used in the next batch again.
For the continuous long-time experiments a horizontal-column packed-bed reactor was used [9, 10]. In this fermenter the gel beads were placed in an eccentrically situated inner column perforated over its entire length. In order to guarantee a free gas release a series of gas outlet tubes were symmetrically inserted in the external column. The reactor had a working volume of 0.140 l and was used as a plug-flow fermenter and as a packed-bed reactor with liquid recycling. In the latter case the recycle rate amounted to 1 l/h. In all of the continuous fermentation experiments the mean hydraulic dilution rate referred to the working volume had a value between 0.28 h^{-1} and 0.35 h^{-1}.

Medium

The feeding-medium for discontinuous fermentations had the following composition:

sucrose	: 136.0 g/l,	K_2SO_4	: 0.58 g/l,
yeast extract	: 5.0 g/l,	$MgSO_4 \cdot 7H_2O$	: 0.22 g/l,
NH_4Cl	: 3.0 g/l,	$FeSO_4 \cdot 7H_2O$	: 11.0 mg/l,
$ZnSO_4 \cdot 7H_2O$	: 6.7 mg/l,	$MnSO_4 \cdot 5H_2O$	: 6.7 mg/l,
H_3PO_4 (85%)	: 0.6 ml/l.		

$CaCl_2 \cdot 6H_2O$ (4.4 g/l) was added to the feeding-medium if alginate gel was used as immobilization matrix. In experiments with pectate-entrapped cells a supply of Ca^{++} was not necessary. For continuous fermentations the medium contained a mixture of molasses (150 g/l) and sucrose (60—110 g/l) as substrate.

Analytic Methods

The biomass content of immobilizates was indirectly estimated by measuring the dry weight of gel beads before and after fermentation. Moreover, the biomass immobilized in alginate gel was determined gravimetrically after dissolving the beads in phosphate buffer. The cell number and cell viability were ascertained in the manner described previously in [11], where the pectate gel beads were destroyed by means of a mechanical ultradisintegrator. The ethanol concentration in the liquid phase was measured by gas chromatography and as an auxiliary criterion for ethanol formation the volumetric CO_2 formation rate was also estimated.

Results

In order to compare both immobilization matrices a series of discontinuous fermentations with repeated use of the immobilizates were carried out with Ca alginate entrapped and Ca pectate-entrapped cells under the same process conditions. Two of the batch curves obtained with small beads of the two gels are shown in Fig. 1. There are no significant differences regarding the course of fermentation between the two variants. Analogous conclusions may also be derived from Tab. 1 which presents more detailed results of this research.

Tab. 1: Results of discontinuous ethanolic fermentation with cells of *Saccharomyces cerevisiae* IBT H 191 entrapped in Ca pectate and Ca alginate gel beads

Parameter	Gel		
	Ca pectate		Ca alginate
Diameter of gel beads before cultivation [mm]	1.44	2.65	1.10
Diameter of gel beads after 9 batch cycles [mm]	1.56	2.75	1.45
Gel volume increase [%]	27	12	129
Biomass load of gel after 6 batch cycles [g dry wt. cells/kg gel]	152	155	154
Mean specific ethanol formation rate of immobilized cells [g ethanol/g dry wt. cells · h]*	0.68	0.46	0.70

* measured on the linear range of the conversion-time curve

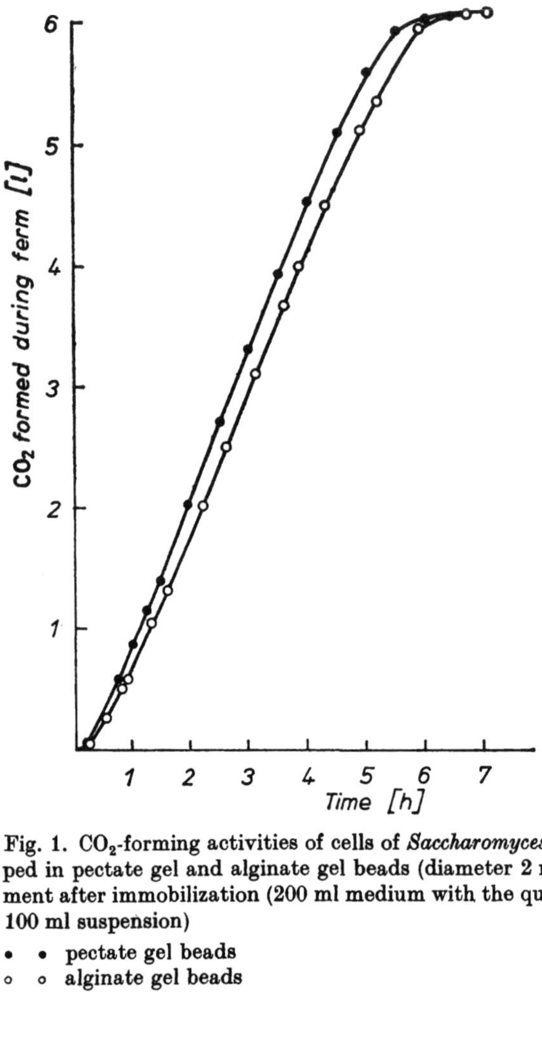

Fig. 1. CO_2-forming activities of cells of *Saccharomyces cerevisiae* IBT H 191 entrapped in pectate gel and alginate gel beads (diameter 2 mm) in the first batch experiment after immobilization (200 ml medium with the quantity of beads obtained from 100 ml suspension)

• • pectate gel beads
o o alginate gel beads

Firstly, we can point out that the biomass loading capacity is equal for alginate and pectate gels. Indeed, the value of 152—155 g dry wt. biomass/kg gel obtained after 6 consecutive batch cycles in each case seems to be the maximum biomass loading concentration for the gels used. No further increase of the cell concentration in the beads was observed after continuation of the repeated batch technique. In this case the amount of biomass in the gel remained constant, whereas the dimensions of cell escape from the gel into the gel-external liquid increased. Secondly, the correspondence of the mean specific ethanol formation rates observed in experiments with small beads from the two gels is clearly shown. The small difference between the measured values of 0.68 and 0.70 g/gh is within the limits of error of the estimation method. But the fermentative performance of cells entrapped in pectate gel can significantly be impaired by diffusion resistances, what could be demonstrated in fermentations with greater beads.

There are significant differences in regard to the swelling behaviour of the two gels. For small beads with a diameter of 1.1 to 1.5 mm the volume increase after 9 batch cycles amounted to 129% for alginate gel and only 27% for pectate gel. The tendency observed in fermentations with Ca alginate-immobilized yeast cells that the swollen beads split into two or more parts of about the same size was found to be less typical of pectate-entrapped cell cultivation. The pectate gel is more brittle than the alginate gel and therefore pectate beads with maximum load are less resistant to shearing effects than alginate beads. Particularly greater beads were quickly destroyed under strong stirring. Under conditions of high fermentative activity in a CSTR beads with maximum load of a diameter of 2.65 mm showed a weight loss of 14% after 15 h compared to 4% in the case of small beads. On the other hand also pectate beads with a diameter of > 3 mm kept their stability over several weeks if mechanical stress could be eliminated, as by using a horizontal column reactor (Tab. 1).

A mean ethanol concentration of 76.0 g/l was measured at the fermenter output during a time interval of 140 h if small beads (diameter: 1.7 mm) were used and the horizontal-column fermenter operated as a plug-flow reactor. As observed in the comparison experiment with alginate-entrapped cells a high number of cells escaped mainly from the portion of gel placed in the first third of the column, where the ethanol concentration was so low that the inhibition of growth was negligible. These free cells multiplied in the reactor and settled between the gel beads with the result that the flow within the column was hindered and the productivity of ethanol formation decreased significantly. Therefore the process had to be finished after 140 h in this case. In the comparison experiment with alginate-immobilized cells this effect was enhanced by the considerable gel volume increase. Here a distinct loss of productivity occurred already after 24 h (Fig. 2).

For overcoming this problem the subsequent experiments were carried out in such a way that the horizontal-column reactor was coupled with a liquid recycle system. So the un-

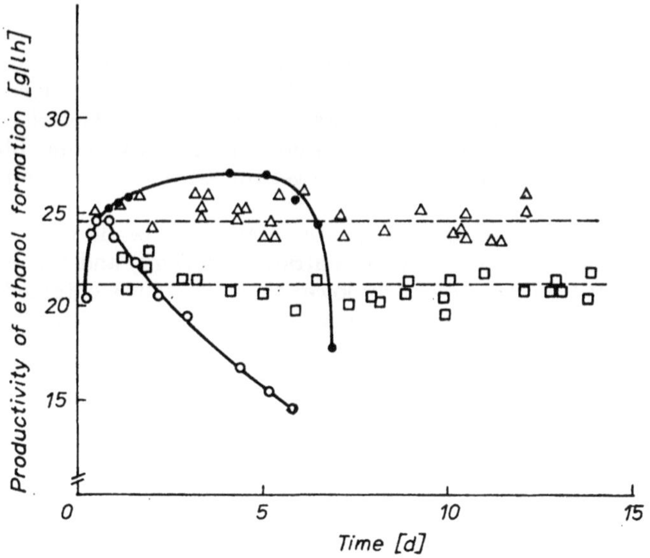

Fig. 2. Productivities of ethanol formation in continuous fermentations performed in a horizontal-column packed-bed reactor with and without liquid recycle using cells of *Saccharomyces cerevisiae* IBT H 191 entrapped in Ca pectate and Ca alginate

- • • pectate gel beads, without liquid recycle
- ○ ○ alginate gel beads, without liquid recycle
- △ △ small pectate beads (diameter: 1.8 mm), with liquid recycle
- □ □ great pectate gel beads (diameter: 3.2 mm), with liquid recycle

desirable growth of the free cells was largely repressed by the stronger product inhibition acting over the entire reactor volume.
Some continuous fermentation experiments were performed in a CSTR using pectate- and alginate-entrapped yeast cells. The gel-specific activities obtained were nearly equal for both immobilizates in each case. The maximum values obtained under conditions of negligible product inhibition (ethanol concentration: 20 g/l) amounted to 103.0 g ethanol/kg gel · h and 108.5 g ethanol/kg gel · h for pectate and alginate gel beads, respectively.
The results of long-time experiments carried out with pectate beads in a horizontal-column reactor are shown in Tab. 2. Nevertheless, the viability of immobilized cells was

Tab. 2: Results of continuous ethanolic fermentation with pectate-entrapped cells of *Saccharomyces cerevisiae* IBT H 191 cultivated in a horizontal-column reactor with and without liquid recycling

Diameter of beads [mm]	Duration of fermentation [d]	Liquid recycle rate [l/h]	Hydraulic flow rate [ml/h]	Ethanol concentration [g/l]	Productivity [g/lh]*	Viability [%] inlet	outlet
1.7	7	0	46.3 ± 2.8	76.0 ± 4.4**	25.0 ± 1.8**	59	19
1.8	12	1	50.2 ± 1.6	69.1 ± 4.2	24.7 ± 1.5	87	69
3.2	13	1	48.9 ± 1.5	60.7 ± 2.5	21.2 ± 1.0	98	83

* referred to the working volume
** for a duration of 5.8 d only

higher than that found in experiments without liquid recycling. As the main reason for this the better availability of nutrients for all of the cells in the well mixed system may be assumed. Under conditions of liquid recycling a mean productivity of 24.7 g/lh could be kept constant over a much longer period at a mean hydraulic dilution rate of 0.35 h^{-1}. The ethanol forming performance was yet the same as at beginning on process interrupting after 12 d.
Similiar results were also obtained with greater beads. For beads with a diameter of 3.2 mm the mean steady-state ethanol concentration (60.7 g/l) and the corresponding mean productivity (21.0 g/lh) were clearly lower than the values obtained analogously with small beads, but a good process stability was yet observed after 13 d also in this case.

Conclusions

As an immobilization matrix for ethanol forming yeasts the pectate gel is a true alternative to alginate. There are no essential differences between the two gels regarding the immobilization procedure and the maximum biomass loading of the gels. The correspondence in the specific ethanol formation rates of yeast cells immobilized in small beads of the two gels and the fact observed previously [12] that alginate-entrapped (diameter of beads: 1.6 mm) and free yeast cells showed the same metabolic performance under analogous conditions signify that no matrix-specific diffusion hindrances were present in the fermentation system used. In contrast in fermentations with greater pectate gel beads (diameter: 2.65 mm) a distinct diffusion resistance was the cause for a mean specific ethanol formation rate which was lower by one third (Tab. 1).

Immobilizates from pectate gel can be applied advantageously in reactors characterized by a low level of mechanical stress and a sufficient intermixture of the liquid phase. The horizontal column reactor with liquid recycling represents such a suitable fermenter system. The long-time stability of pectate gel beads tested in this reactor and the fermentative performance of the so immobilized cells were very good. Because of their lower swelling capacity and hence smaller tendency for bed plugging in packed-bed reactors pectate gel beads were superior to alginate gel beads in long-time fermentation experiments.

Received February 23, 1989

References

[1] Vorlop, K.-D.: Thesis, TU Braunschweig, 1984.
[2] Navarro, A. R., Rubio, M. C., Callieri, D. A. S.: Eur. J. Appl. Microbiol. Biotechnol. **17** (1983), 148.
[3] Navarro, A. R., Marangoni, H., Plaza, I. M., Callieri, D.: Biotechnol. Lett. **6** (1984), 465.
[4] Noguchi, S., Nagashima, M., Azuma, M., Furukawa, S.: CA **103** (1985), 1191098.
[5] Noguchi, S., Nagashima, M., Azuma, M., Furukawa, S.: CA **104** (1986), 33052.
[6] Anon. JP 8313387, 8313391 (1983), CA 98 (1983), 196384, 214167.
[7] Berger, R., Rühlemann, I., Maune, Richter, K., Steudel, A.: DD 258419 (1987).
[8] Berger, R., Rühlemann, I.: Acta Biotechnol. 8 (1988), 401.
[9] Berger, R., Rühlemann, I.: DD 221470 (1985).
[10] Rühlemann, I., Berger, R.: Proc. third Colloquy on Research Equipment for Biotechnology, Heiligenstadt, G.D.R., November 3–6, 1986, 191–198.
[11] Berger, R., Rühlemann, I.: Acta Biotechnol. 8 (1988), 395.
[12] Richter, K., Rühlemann, I., Becker, U., Berger, R.: Acta Biotechnol. **9** (1989), 123.

Acta Biotechnol. 10 (1990) 1, 62 Akademie-Verlag Berlin

Book Review

I. Hargittai, M. Hargittai

Symmetry through the Eyes of a Chemist

New York: VCH Publishers, Inc.

458 pp., 380 photographs, figures and drawings

Today it is a commonplace belief that, not only is symmetry one of the fundamental concepts in science, but it is also possibly the most significant concept bridging numerous branches of the sciences, the arts and many other human activities.

The Hargittai's wrote an unusual book. It reveals connections between seemingly unrelated things: scientific findings, everyday phenomena, creations of art and literature. The authors survey chemistry from the viewpoint of symmetry, a perspective which yields an understanding of molecular geometry and the structure of crystals.

After the introduction (Chapter 1), the simplest symmetries are presented using chemical and non-chemical examples (Ch. 2). Molecular geometry is then discussed in qualitative terms (Ch. 3). Group theoretical methods (Ch. 4) are applied in an introductory manner to the symmetries of molecular vibrations (Ch. 5), electronic structure (Ch. 6), and chemical reactions (Ch. 7). These chapters are followed by a descriptive discussion of space-group symmetries (Ch. 8) including the symmetries of crystals (Ch. 9).

The discussion of molecular vibrations, electronic structure, and chemical reactions requires some mathematics, which is fully provided within the book. Lucid and striking illustrations help to grasp abstract ideas. With their approach the authors place chemistry into a broader context, enabling readers to perceive more than meets the eye in the world around us.

Symmetry is also inherent in animate nature — both at a macrolevel (plants, animals, microorganisms) and at a microlevel of the structure of biomolecules and their various associations. As an example, the phenomenon of chirality appearing in constituents of living organisms and probably being one of the clue problems relevant to the origin of life is discussed.

In his foreword to a special issue of **Computers and Mathematics with Applications**, Vol. 12, No. 1/2 "Symmetry Unifying Human Understanding" (which should be named in the same breath as the book to be reviewed), Erwin Y. Rodin said "If indeed symmetry plays such an important role in so many human endeavors and phenomena, and if it may be one of the clues to a better understanding of where, what and who we are, we ought to speak more about this subject". Therefore, it is a book not only for chemists. The circle of readers is hardly to delimit other than saying the book is dedicated (and recommended) to all those who are interested in considering integrating concepts and principles in life sciences.

B. Nagel

Acta Biotechnol. 10 (1990) 1, 63–71 Akademie-Verlag Berlin

FERM — a Decision Support System for Fermentation Processes

GUTHKE, R.[1], SCHULZ, V.[1], KNORRE, W. A.[1], HAUPT, B.[2], WERNSTEDT, J.[2]

[1] Academy of Sciences of the German Democratic Republic,
Central Institute of Microbiology and Experimental Therapy,
Beutenbergstr. 11, Jena, 6900 G.D.R.
[2] Technical University Ilmenau,
Department of Technical and Biomedical Cybernetics,
Ilmenau, 6300 G.D.R.

Summary

For recombinant DNA product fermentations in the laboratory and pilot scale as well as for antibiotic fermentations at the industrial scale decision support program systems for personal computers were developed. It assists the operator to solve control and decision tasks. Applying this program the data and knowledge bases increase, which include measured and derived process data, parameters and comments of previous fermentation runs as well as algorithmic and declarative rules.

Introduction

During and in preparation of a fermentation run one has to solve a lot of decisions and control tasks in keeping the qualitative and quantitative goals. Hitherto in industrial practice such tasks could be solved successfully for partial processes only, for instance the control of sterilization. But the fermentation process as a whole can not be controlled automatically due to the complexity and incompleteness of the available process information. Thus, in the practice the fermentation is supervised empirically where the man with his limited experiences and subjective insufficiencies makes the decisions. Decision support or expert systems become favourable tools to overcome these problems [1–6]. With the help of computers such a system provides to the operator a set of proposals for the solution of control and decision tasks on the basis of knowledge collected from various specialists and selected according to the current process state. The given proposals on control tasks are explained and interpreted by the computer system to the operator. Finally, if he is authorized, the operator makes himself the decisions on the basis of these proposals and his own experience and observations. After acquiring the results of fermentation, for instance when the product yield and costs are obtained, the decision support or expert system estimates the quality of the realized decisions to change the knowledge basis or to force the man in learning.
This paper presents the structure of and first experiences with a decision support system (or "mini-expert system") developed for antibiotics and recombinant-DNA-product fermentations.

Materials and Methods

For regular application of the decision support system in fermentation control GDR made personal computers were used, for instance the CP/M 86-compatible type A-7100 (VEB ROBOTRON, GDR). The software was written in the language Turbo-PASCAL. For comparision with established tools of artificial intelligence a software developement using PROLOG was started.
At present the decision support system for fermentations, called FERM, is realized in versions for the laboratory, pilot and industrial scale:

— The version for laboratory scale works in on-line coupling with the laboratory fermenter LFS-212 (MYTRON Heiligenstadt, GDR). Here the on-line available data, such as pH, pO_2, O_2, CO_2, dosage pulse number, netto weight, temperature, aeration rate, impeller spead, power input, and the biomass sensor output, are sent to the personal computer via the current loop output channel of the computerized fermenter to the personal computer where the data are buffered. From buffer the data are read during the input loops of menue procedures and are stored at diskettes.

— The version for pilot scale works at present off-line, only. But a data transfer is possible from microcomputer K-1520 coupled via CAMAC-interface with the pilot fermenter CHEMAP-450 [7].

— Also the version for industrial scale works at present offline, only. This version is developed for the high-equipped and computerized 63-m^3-fermentation-plant in the factory VEB JENAPHARM (Jena, GDR).

Structure of the Decision Support System

The decision support system FERM consists of the following parts:

— input and output modules, which work in dialog,
— data and knowledge basis, and
— decision module [1].

Input/Output Module

The acquisition of off-line available informations, the output of decision and control proposals as well as the presentation of explaining knowledge and diagnostic informations on the current state are performed by man-machine-dialog. For this dialog the output to the operator is given on an alphanumerical-graphical display and on request at a printer. The input from the operator to the decision support system is possible by a keybord. In this dialog modules a set of options are given as a menue to the man who selects the desired or default option.

Data and Knowledge Basis

Informations, i.e. the data and knowledge basis, are stored in various files on diskettes. These informations are structured in rules and facts where the

— rules are algorithmic or declarative ones and the
— facts are parameters, mesaured data and comments.

The *algorithmic knowledge* (for instance equations and models for simulation) is given by procedures and functions in Turbo-PASCAL-text-files or by compiled program modules which can be called. The *declarative knowledge* (rules, IF-THEN statements and verbal informations) is presented in structured text files which can be written, edited, copied, and printed using comfortable text processors. Applying the decision support system for fermentations in the scientific research it was found that it is favourable to create for

the declarative knowledge two text files or two groups of text files, a constant and a variable one. In a constant file for instance the instruction for the estimation of ammonia nitrogen by Kjeldahl's method was included. In a variable file were written for instance instructions for the preparation technology of complex media or for the inoculation.
The facts of the knowledge basis are stored in three files or groups of data or text files. The first contains the data measured on- or off-line. It is favourable to store the on-line *measured data* as integer or byte numbers whereas the off-line measured data as real numbers or in a text file. The information on the structure of these data files was encoded by a second type of data, the *parameters*. Such a parameter is a complex datum and consists of numerical and verbal informations:

— the symbol (given by two or more characters),
— the name (given for instance by 20 characters),
— the numerical value (given by a real number),
— the dimension (given by four or more characters).

In this type of facts a lot of other informations are stored, such as

— the composition of main and preculture media,
— prices of the ingredients,
— technical parameters of pumps and sensors,
— set points and operation parameters for controllers

and other cultivation parameters.
This sets of parameters are structured in groups labelled by the nomenclature of symbols.
The third type of facts is the verbal information containing the *comments*. These comments inform about the disturbances during the process and about other events. They are related to a certain time point of the fermentation run and can be stored as a separate text file or together with the off-line measured data file.
All these files can be edited with the help of either a text processor (for text files) of by other tools included in the decision support system (for numerical files). Creation, edition and storage of these facts can be done during the running process by an authorized operator after input of a code-name used for secrecy. The input of new measured data and comments are permitted within a time window for everybody. After this input the facts are stored at diskettes. When starting a new fermentation run an authorized operator can create both a new parameter file and a new variable text file for declarative knowledge by making a copy of a selected old file and by modifications. This is a very simple mode of learning. As an advanced mode of learning process specific and model based modules of data analysis have been developed and being under work for adaptation and optimization of certain process parameters.

Decision Module

The purpose of the decision support system is to give decision and control proposals on the basis of both, actual informations on the running process as well as using the stored knowledge. These proposals are presented and explained to the man who makes the decisions taking in account both, these proposals as well as his own experiences. The decision module has to solve two tasks:

1. recognition and classification of the actual situation and
2. transformation of the classified situation to an appropriate decision or control proposal.

First of all the situation will be specified in a man-machine-dialog by a tree of simple decision options. For instance in the main menue one has to select one of the following options:

- preparation (before the fermentation is running),
- execution (when the fermentation is running),
- analysis and final report (when the fermentation was finished).

As the default option the escape is used.
In the phase *preparation* of fermentation a set of instructions for both biological and technical manipulations as well as informations about the scientific or economic aim are presented. Here informations about

- the microorganism used,
- the conditions of its conservation and preculture,
- the technology of preparing the media (succession of addition of ingredients, hydrolysation or enzymatic pretreatment of complex components if any, temperature and pressure profile during sterilization, pH-handling),
- the preparation of analytical methods and preparations of sensors,
- the preparation of dosage equipments and various controllers,
- and other

are offered in verbal form together with quantitative informations stored as parameters (see above). Here these facts and rules can also be modified or listed on request by authorized operators.
The phase of *execution* of fermentation run starts with a check list to verificate the correct state of the fermenter and its peripheric device. Then a menue is offered with the following options:

- data acquisition (off-line),
- data analysis,
- decision and control proposals,
- data presentation of kinetic data in tables and graphics,
- instructions.

In the regular application the data acquisition is the standard option which is followed by the options data analysis and presentation of decision and control proposals. The *data acquisition* assists the input of off-line measured data together with its transformations. Here the identification and adaptation of calibration functions used for transformations are included. For instance

- the estimation of phosphate concentration uses a linear relation fitted by linear regression,
- the biomass concentration is estimated from extinction as well as scattered light intensity using nonlinear functions, and
- the product concentration (antibiotics as well as interferon) is determined by bioassays in connection with nonlinear calibration functions.

Together with off-line data comments can be stored, too. An operator who is authorized by a code-name can also cancel, correct and add data or comments subsequently on request.
The number of measured data which are resident in the memory differs from version to version. In the version for industrial antibiotics fermentation 200 data sets each with 25 values are resident.

After the acquisition of data and other informations the *data analysis* is called to recognize and classificate the actual state of the running process. Derived informations are calculated, such as

- the absolute or specific (i.e. biomass related) substrate consumption rates, for instance of the carbon source (glucose or starch), ammonia-nitrogen and oxygen,
- the productivity and specific product formation rate for the desired product and byproducts (CO_2),
- the absolute and specific (i.e. product yield related) process costs,
- the specific growth rate calculated from biomass concentration determined from different observables (dry weight, extinction, scattered light intensity, heat and CO_2-production or energy and carbon balance),
- yield coefficient for selected substrates,
- oxygen and heat transfer coefficients.

Algorithms for situation recognition by classificators are under work now. For instance methods of fuzzy process analysis were successfully proven to classify the relations between specific growth rate and specific rate for recombinant-DNA-derived product formation [8]. The goal of such classification is to maximize the specific product formation rate by optimal control of the specific growth rate via optimal substrate feeding.
In the established regular application algorithms are used which permit proposals on the substrate (glucose or starch and nitrogen) feeding to improve the control reliability for a desired (low) level of substrate concentration in the fermentation broth or in the outflowing medium. This decision module is very important especially if the process is performed semicontinuously in repeated draw-off systems as commonly applied for antibiotics production. Thus, this module is the essential one in the large scale version of the decision support system. Fig. 1 shows the hardcopy of a typical feeding proposal obtained for an antibiotics fermentation in the industrial scale.

Decision support system FERM
Fermentation run: NTC R/06/88 Version: 01. 02. 88

Decision Proposal

date	(TT.MM)	= 26.02 1988
hour	(hh.mm)	= 11.00 a.m.

No ammonia-sulfate-dosage
required for the next 24 hours.

Starch-dosage required after 7.0 hours.
Preparations must be started after 1.0 hour.
Proposals for the required amount
desired (Y/N)?

Fig. 1. Example of a decision proposal given by FERM to the operator during the nourseothricin (NTC) production

For repeated fed-batch operation the decision module is formulated more complex by a lot of conditional statements [9]. An example of such a statement for cyclic operation is shown in Fig. 2.
For each fermentation a proposal for ending the fermentation run is important. The condition for finishing depends on the desired performance criterion. In the industrial

```
IF      (       Prz [i] < Prz [i-1]                )
{       Productivity decreases                              }
 AND (          pO2 > pO2_min                      )
        Oxygen supply is not disturbed                      }
AND  (          S_Glucose < S_Glucose_crit         )
{       Substrate dosage is not disturbed                   }
AND  ( (            x >= x_max                     )
{       Biomass measured as sediment is high                }
        OR (         (          pO2 <= pO2_crit    )
{       Dissolved oxygen concentration is low               }
                AND    (      rpm >= rpm_max       )
                AND    (          Air >= Air_max   ) )
{       Oxygen supplay is at the limit                      }
        OR (          SO [i] > SO [i-1]            ) )
{       Limiting substrate concentration
        in the substitution medium
        was increased already                               }
 AND (          Gamma [i] < 0.5                      )
{       Substitution portion Gamma
                can be enlarged                             }
THEN            Gamma [i+1] := Gamma [i] + Delta;
        Enlarge the substitution portion!                   }
```

Fig. 2. Example of a conditional statement in the decision module of the decision support system for the semicontinuous antibiotics fermentation

scale version of the decision support system FERM both the minimization of specific (product yield related) process costs or the maximization of productivity are considered. If not one unique criterion can be selected or formulated by combination then an efficient set of solutions is provided where the operator has to select in manmachine dialog the desired operation mode. In this selection also quality parameters (e.g. refering a low level of byproducts and residual substrates) and restrictions refering the operation capacity of the down-stream processing must be considered for the decision. A module that assists this complex decision on ending of fermentation is under work for the industrial scale.

For the pilot scale a decision module was developed that improves the process performance in respect to the controlled carbon source feeding in connection with agitation control [10]. The decision module acts as a supervisor to the process controller as done analog by Cooney et al. [6]. One task of this supervisor is to switch on the glucose feeding controller when carbon source limitation was recognized.

The decision support system has not only to provide the decision and control proposals but it has to explain and interprete these. The explanations must be presented according to the operator's knowledge using man-machine-dialog. Besides the short explanation as regularly given for each proposal one can call the various data analysis procedures (see above) as well as the presentation of the kinetic process state by tables and graphics. These routines form the *explanatory module*. For the report hardcopies with legends are permitted especially for the graphic representation of the kinetics selected by the operator.

The available knowledge in the form of conditional statements (IF-THEN-relations) as a prerequisite for an expert system, which transform the classified situation to appropriate decision or control proposals, is very small up to now. The elements for fault diagnosis and management will be supplemented step by step after occurance of abnormal and disturbed situations. But for undisturbed situations the established "art of fermentation" is given widely by a set of instructions. Such instructions are offered about

- the sampling mode and the pretreatment of sampled material,
- repeated calibration of sensors
  such as for biomass measurement by light scattering or the exit gas analysis (O_2, CO_2),
- the analytical methods for biological and chemical observables, such as for product, biomass, glucose, nitrogen and phosphate concentration,
- the goal of the fermentation run.

This instructions are valuable for the interdisciplinary cooperation in the laboratory or for standardization in the fermentation plant.

Discussion: First Experiences with FERM

In the Institute ZIMET Jena were and will be developed various versions of the decision support system FERM for the process development for the production of recombinant-DNA-derived products such as interferons alpha-1 and alpha-2 in the laboratory and pilot scale and for the optimization and stabilization of industrial nourseothricin [11] production in the factory VEB Jenapharm. In both cases only a few rules for the decision making are available. The extension of this knowledge is a prerequisite for building up an expert system. For this a lot of procedures are developed and tested for a-posteriory *data analysis* to find out more conditional statements between the recognized process state and the optimal manipulations.
Some of the procedures regard with the full or selected set of previous fermentation runs which data are stored on diskettes. These procedure use statistics and permit various tests and the print of histograms and correlation functions. Fig. 3 shows an example for data analysis in respect to growth and product formation kinetics of an ensemble of fermentation runs. 10 comparable batch fermentations in a 3-liter-laboratory fermenter

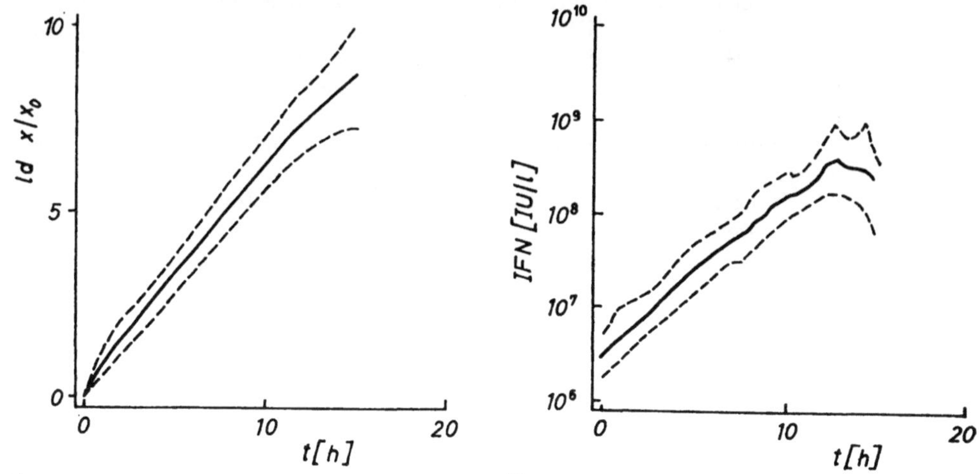

Fig. 3. Elements of the knowledge basis: a "standard course of fermentation" with interferon-alpha-2 producing genetically engineered *Bacillus subtilis* calculated as arithmetic mean and standard deviation of 10 fermentation runs in a 3-liter laboratory fermenter

a) Growth kinetics in semi-logarithmic plot
b) Product formation kinetics in semi-logarithmic plot

(LF-2 from MYTRON Heiligenstadt, GDR) with glucose-mineral-salt medium inoculated with a genetically engineered *Bacillus subtilis* [8] are selected from about 50 runs in this example. The measured data for biomass as well as interferon-alpha-2 (as the fermentation product) concentration are linearly interpolated. Then for each time point the arithmetic mean together with the standard deviation was calculated by averaging of the 10 interpolated kinetics. In doing so one gets a "standard course of fermentation" as one element for the knowledge basis. This knowledge provides the basis for the classification of a running fermentation as normal or abnormal.
Other procedures of data analysis take into account one selected fermentation run only analog to the data analysis tools for the execution phase as described above. But here also algorithms are included which analyse the full fermentation kinetics from the start up to the end. Applying methods of piecewise linear regression which take into account a given error of the measured observable one can recognize the different phases of growth and product formation. Alternatively methods of smoothing cubic splines are used for calculation of quasicontinuous kinetics of derived parameters such as the specific growth and product formation rates and the differential yield coefficients for which the first and second derivation to the time is necessary [12]. By dialog the error of the variable or the smoothing factor, respectively, can be variated and the appropriate one can be choosen for regulare use to

— recognize the process phases during the future fermentation runs and to
— test the consistency of kinetic data.

These procedures of data analysis are proven to be valuable for a fermenter supervisory control. These together with routines for the report, i.e. the table print of — full or selected — measured data, parameters and comments, can be called via the main menue of the decision support system. The report routines are appropriate tools for data and knowledge representation which replace step by step the manual process documentation.
The efficiency and reliability of kinetic data analysis and process documentation could be improved significantly using FERM. The improvement of process performance and control reliability as the main goal of an expert system were inconsiderable up to now. The core of the decision support system increases with the knowledge basis that is still rudimentary. When grown up the knowledge basis the decision support system or expert system must be implemented in more powerful computers including artificial intelligence technology.
The ability of self-learning as an essential feature of expert systems is realized in the decision support system FERM only weak. A few parameters such as the yield coefficients can be adapted from fermentation to fermentation as well as during the running process where proposals for manual adaptation by the operator are provided by the computer program on the basis of statistic tests. But the self-learning of rules is not realized by the system yet. At the current stage the rules are learned by the scientist who analyses the process data and who programmes the found rules in a program-module using the comfortable Turbo-PASCAL-editor. This is tolerable as long as the knowledge basis is small and not fast growing.

Received January 20, 1989

References

[1] Wernstedt, J.: msr 28 (1985), 295.
[2] Foxboro: Verfahrenstechnik 10 (1986), 25.
[3] Angelov, A. I., Denčeva, E. C., Minkova, H. A.: Bioavtomatika (Sofia) 2 (1986) 59.

[4] Endo, I., Nagamune, T.: Bioprocess Eng. **2** (1987), 111.
[5] Beltramini, L., Motard, R. L.: Chem. Biochem. Eng. Q. **2** (1988), 119.
[6] Cooney, C. L., O'Connor, G. M., Sanchez-Riera, F. — In: Durand, G., Bobichon. L., Florent, J. (Eds.) Proc. 8th Int. Biotechnol. Symp., Paris, 1988, 563.
[7] Veith, G., Zuber, G., Schumann, K.: Wissenschaftliche Beiträge der IHS Köthen **85** (1985), 52.
[8] Guthke, R., Peissker, M., Tonew, M., Glück, B., Votruba, J.: Submitted to Chemical and Biochemical Engineering Quarterly.
[9] Guthke, R., Peissker, M., Gira, G., Hess, W., Menner, M., Mund, K., Knorre, W. A.: WP-DD 252 002 A1, 1986.
[10] Riesenberg, D., Menzel, K.-D., Schulz, V., Günther, J., Gira, G., Knorre, W. A.: WP-DD 318370 5, 1987.
[11] Bocker, H., Bergter, F.: Arch. Exp. Veterinaermed. Leipzig **40** (1986), 646.
[12] Reimann, B., Mund, K.: Z. Allg. Mikrobiol. **24** (1984), 485.

Book Review

R. Crespi

Patents: A Basic Guide to Patenting in Biotechnology

(Cambridge Studies in Biotechnology: 6)

Cambridge: Cambridge University Press, 1988.
191 S., $ 49.50, ISBN 0521-32954-X

Das vorliegende Buch stellt einen Grundriß des Patentrechts für Wissenschaftler insbesondere der Fachrichtung Biotechnologie dar. Der Autor, einer der international führenden Experten auf dem Gebiet des Rechtsschutzes für geistiges Eigentum, war früher selbst viele Jahre auf diesem Sektor in der Industrie tätig und kennt die Probleme aus erster Hand, vor die Biotechnologen gestellt sind, wenn sie mit Patentrechtsfragen in Berührung kommen. Das vorliegende Buch ist bestens geeignet, diese Materie für den Wissenschaftler handhabbarer zu machen.
Crespi behandelt zunächst allgemeine Grundfragen des Patentrechts und macht den Leser anschließend mit dem Ablauf des eigentlichen Patentierungsprozesses vertraut. Nach einer ausführlichen Abhandlung der Schutzvoraussetzungen wird auf Besonderheiten des Rechtsschutzes für biotechnologische Erfindungen eingegangen, wobei die „neue Biotechnologie" wie auch der Schutz von Ergebnissen der Pflanzen- und Tierzüchtung gesondert berücksichtigt werden.
Der Leser bekommt alle Informationen in Form eines Frage-Antwort-Spiels vermittelt. Der Autor hat dabei alle Problemkreise sorgfältig abgedeckt, die für einen Biotechnologie-Fachmann im Zusammenhang der Publizierung, andererseits mit dem Patentschutz seiner Ergebnisse auftreten könnten.
Etwaige Unterschiede nationaler Bestimmungen werden insbesondere für USA, Großbritannien und Bundesrepublik Deutschland berücksichtigt; ansonsten geht Crespi von den rechtlichen Regelungen des Europäischen Patentübereinkommens aus. Für den Biotechnologen aus diesem Rechtsbereich sollte das Buch zum festen Bestand seiner Handbibliothek gehören; über diesen Personenkreis hinaus ist es überaus nützlich und auch interessant, sich auf die dargebotene Weise in das diffiziele Gebiet des Patentschutzes in der Biotechnologie einführen zu lassen.

B. Konieczny

Acta Biotechnol. 10 (1990) 1, 73—78 Akademie-Verlag Berlin

Effects of Temperature on the Growth of the Thermotolerant Methylotrophic Bacillus sp. NCIB 12522 in Chemostat Culture

AL-AWADHI, N., HAMER, G.*, EGLI, TH.

Institute of Aquatic Sciences
Swiss Federal Institute of Technology Zürich,
Ueberlandstraße 133, CH-8600 Dübendorf, Switzerland

Summary

The effects of temperature on chemostat cultures of a thermotolerant, methylotrophic *Bacillus* sp. are reported. In spite of the relative fastidiousness of this bacterium, overall response to a range of growth temperatures is predictable. However, anomalous results with respect to biomass yield coefficients are evident, suggesting complex and extensive endogenous and exogenous activities under some growth conditions.

Introduction

The application of continuous flow enrichment techniques has recently allowed the isolation of several thermotolerant, methylotrophic *Bacillus* sp. from aerobic thermophilic bioreactors used for the pretreatment and hygienization of waste sewage sludge [1]. The isolation of such strains was unexpected because of earlier failures to isolate either thermotolerant or thermophilic methylotrophs in extensive screening programmes during the era of single cell protein production from petrochemical feedstocks. The major reason why thermotolerant methylotrophs could not be isolated previously is probably their deactivation very soon after the exhaustion of methanol during batch enrichment cultures [2].
Here, it is sought to examine the effects of temperature on the growth of one of the novel thermotolerant methylotrophic bacilli isolated, i.e., *Bacillus* sp. NCIB 12522, in methanol-limited chemostat culture. The reason for such studies is that it is expected that thermotolerant methylotrophic bacilli will find application in the aerobic biotreatment of hot petrochemical industry wastewater as it cools through the thermotolerant temperature range for microbial growth, i.e., 60° to 45 °C. In batch culture, *Bacillus* sp. NCIB 12522 has been shown to exhibit a maximum specific growth rate constant of $0.69\ h^{-1}$ between 45° and 55 °C, whilst at temperatures below 45 °C, the maximum specific growth rate declined steadily and at temperatures exceeding 55 °C, a precipitous decline was observed [2]. No growth occurred at 60 °C. Such behaviour in batch culture suggests that in chemostat culture wash-out will occur at different temperatures depend-

* To whom correspondence should be addressed.

ing on the dilution rate employed. In order to determine whether wash-out in chemostat culture can be predicted from batch growth rates, the present study was initiated.
In all processes involving the growth of microbes, endogenous metabolism is judged to be a significant factor affecting the overall biomass yield coefficient. Traditionally, endogenous metabolism is explained on the basis of the maintenance requirements of microbes (3). However, in technical-scale processes for either wastewater or waste sludge biotreatment, endogenous metabolism (activity) is much more loosely defined and embraces a series of exogenous processes that are clearly excluded from the traditional definition (4). These include simultaneous "cryptic" growth of the culture on lysis products and growth on any other extracellular products formed. Lysis, death and "cryptic" growth were generally considered to be features of mixed culture systems, but recently, MASON and HAMER [5] demonstrated the lysis/death/"cryptic" growth cycle to be of considerable importance during the growth of a pure monoculture of *Klebsiella pneumoniae*.

Materials and Methods

Organism

Bacillus sp. NCIB 12522, a restricted facultative thermotolerant methylotroph isolated from a technical-scale aerobic thermophilic waste sewage sludge biotreater by AL-AWADHI et al. [1] was used. The bacterium was maintained on 3M methanol agar slopes over 4—5 months at 4 °C. Working stock cultures were maintained on 3M methanol agar plates incubated at 48 °C and transferred every 2 days.

Medium

The *Bacillus* sp. was grown in a defined mineral salts medium which contained per litre: H_3PO_4 (85% solution), 0.1 ml; K_2SO_4, 50 mg; $MgSO_4 \cdot 7H_2O$, 160 mg; NaCl, 50 mg; $CaCl_2 \cdot 2H_2O$, 200 mg; $(NH_4)_2SO_4$, 500 mg; EDTA-Na_2, 50 mg; trace element solution, 1 ml. The trace element solution used comprized per litre: $FeCl_3 \cdot 6H_2O$, 12 g; $CuSO_4 \cdot 5H_2O$, 1.6 g; $ZnSO_4 \cdot 7H_2O$, 2.2 g; $MnSO_4 \cdot 4H_2O$, 1.7 g; $Na_2MoO_4 \cdot 2H_2O$, 300 mg; $CoCl_3 \cdot 6H_2O$, 800 mg; H_3BO_3, 50 mg; HCl (fuming), 35 ml. 1 g l^{-1} methanol was used as the carbon energy substrate and 20 mg l^{-1} polypropylene glycol was added as an antifoam agent.

Bioreactor

The bioreactor used was 3.4 l total volume (Bioengineering AG, Wald, CH). It was fitted with temperature, impeller speed and pH control, dissolved oxygen monitoring and both pumped medium inlet and culture overflow weir so as to allow operation in the continuous flow (chemostat) mode. Both the oxygen and the carbon dioxide content of the effluent gas stream from the bioreactor could be measure on-line. For all cultivation experiments, the culture pH was maintained at 6.8 by automatic addition of an equimolar 0.5 N NaOH/KOH solution. The operating temperature of the biorector was varied from experiment to experiment, the operating volume was 1.5 l, the impeller speed used was 800 rpm and the inlet air flow rate was 14.5 l h^{-1}.

Dry Weight

Dry weights were measured on triplicate 10 ml samples of culture that were filtered through tared 0.22 μm Nuclepore filters and were dried overnight at 105 °C to constant weight.

Methanol

This was measured by using a gas chromatograph type RlA (Shimadzu Corp. Kyoto, J). The column used was 80/120 carbopack B/6.6% carbowax 20M, 2 m × 2 mm i.d. Column temperature was 80 °C and the flow rate of the carrier gas (nitrogen) was 20 ml/min.

Oxygen Consumption and Carbon Dioxide Production

Oxygen and carbon dioxide concentrations in the effluent gas stream from the bioreactor were measured with a paramagnetic oxygen analyser (Oxymat 3, Siemens AG, D) and infrared carbon dioxide analyser (Binos 1, Leybold Heraeus GmbH, Hanau, D), respectively. Rates were then calculated on the basis of an inert gas balance.

Results and Discussion

The results obtained for chemostat operation at a dilution rate of $0.22\ h^{-1}$ and temperatures between 34 °C and 59 °C and those for chemostat operation at a dilution rate of $0.52\ h^{-1}$ and temperatures between 41° and 57 °C are shown in Figs. 1 and 2, respectively. It is clearly evident from Fig. 1 that the steady-state biomass dry weight did

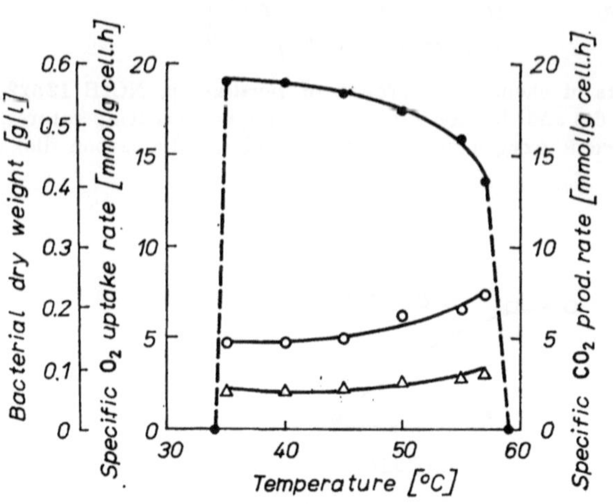

Fig. 1. Results of chemostatic growth of *Bacillus* sp. NCIB 12522 on methanol (1 g/l) at pH 6.8 and dilution rate $0.22\ h^{-1}$ for various temperatures. Biomass dry weight (•), specific oxygen uptake rate (○) and specific carbon dioxide production rate (△)

not remain constant over the temperature range where wash-out was not occurring, indicating a declining biomass yield coefficient between operation at 35 °C and operation at 57 °C. Corresponding increases in the specific oxygen uptake and specific carbon dioxide production rates further support this. In contrast, Fig. 2 indicates an almost constant steady-state biomass dry weight between 42 °C and 55 °C, although the data for the specific oxygen uptake and specific carbon dioxide production rates suggest some reduction in the biomass yield coefficient between 50° and 55 °C. The maximum methanol based biomass yield coefficients observed were 0.57 and 0.56 at dilution rates of $0.22\ h^{-1}$ and $0.52\ h^{-1}$, respectively.

The estimated recovery of methanol carbon in biomass and carbon dioxide is shown for both dilution rates in Fig. 3. It was assumed that dry biomass comprized 50 percent carbon. At the higher dilution rate, recovery is essentially constant at ca. 87 percent irrespective of temperature, suggesting that extracellular product formation was occurring. However, this was not checked. At the lower dilution rate, recovery decreased with increasing temperature. At 35 °C it was estimated to be 96 percent, whilst at 57 °C it was reduced to 86 percent, suggesting that at the lower growth rate either

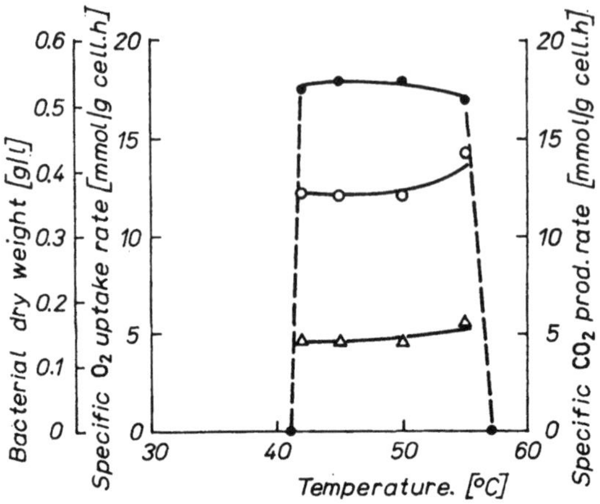

Fig. 2. Results of chemostatic growth of *Bacillus* sp. NCIB 12522 on methanol (1 g/l) at pH 6.8 and dilution rate 0.52 h^{-1} for various temperatures. Biomass dry weight (•), specific oxygen uptake rate (○) and specific carbon dioxide production rate (△)

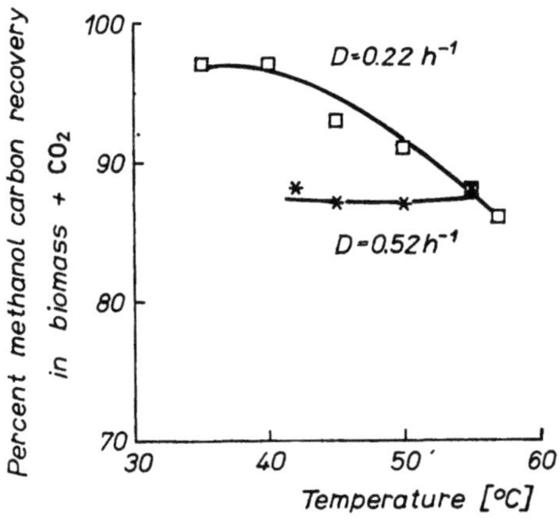

Fig. 3. Calculated values of percent methanol carbon recovery in biomass and CO_2 at dilution rates 0.22 h^{-1} and 0.52 h^{-1} for various temperatures

extracellular product formation was less pronounced or reutilization of product occurred at lower growth tempertures. The deficiency in the carbon balances could not result from changing cell composition as a result of intracellular storage product formation, because such storage products would not markedly affect the percentage carbon content of the cells [6]. Additionally, they would be unexpected under carbon limited growth conditions. In a recent report on the growth of the thermophile, *Bacillus stearothermophilus*, in glucose-limited chemostat culture, PENNOCK and TEMPEST [7] observed significant acetate formation.

Wash-out at a dilution rate of 0.52 h^{-1} occurred at above 55 °C and below 42 °C, whilst at a dilution rate of 0.22 h^{-1}, it occurred above 57 °C and below 35 °C. Variation in the temperatures under which wash-out occurs can be predicted from the batch experiments

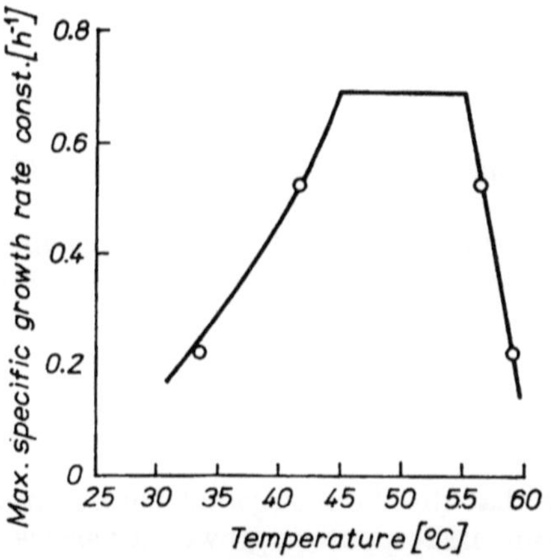

Fig. 4. Relationship between temperature and the maximum specific growth rate constant for batch growth of *Bacillus* sp. NCIB 12522 (2) with superimposed values of critical dilution rates obtained in chemostat culture (○)

reported elsewhere, as is evident from Fig. 4, where the critical dilution rates for steady state continuous cultures are imposed on the relationship between temperature and the specific growth rate constant obtained in batch culture [2]. However, the marked decline in the biomass yield coefficient at a dilution rate of $0.22\ h^{-1}$ when compared with only a minor decline in the biomass yield coefficient at a dilution rate of $0.52\ h^{-1}$, under similar operating temperatures, requires explanation.

The observed decline in the biomass yield coefficient at a dilution rate of $0.22\ h^{-1}$, within and above the optimum temperature range for growth of the culture is consistent with the hypothesis that suggests extracellular product reutilization under some growth conditions rather than with hypothesis for either maintenance requirements or "cryptic" growth. Such extracellular products could, at elevated temperatures, result from cell leakage.

MAINZER and HEMPFLING [8] have attributed reductions in observed biomass yield coefficients with increasing temperature for *E. coli* to result from increasingly uncoupled growth and energy wastage. Earlier, TOPIWALA and SINCLAIR [9] proposed and verified that the endogenous rate constant for *Aerobacter aerogenes* (*K. pneumoniae*) is a function of temperature, but independent of the imposed dilution rate in chemostat cultures. However, their definition of endogenous activity was not restricted to strict maintenance requirements.

MATSCHÉ and ANDREWS [10] have investigated the growth of a *Bacillus* sp. in the thermotolerant and thermophilic temperature ranges for growth and suggest, on the basis of their results, that both maintenance requirements and bacteria decay are enhanced with increasing temperature, although the optimum temperature for the growth of their bacterium was not defined. More recently, KUHN et al. [11] have reported similar results for the thermophilic bacterium, *Bacillus caldotenax*, and have claimed that maintenance requirements for this bacterium in the thermophilic range for growth can be some 10 × those observed for mesophiles growing at mesophilic growth temperatures. However, neither report makes a clear differentiation between endogenous and exogenous events.

Investigations by PENNOCK and TEMPEST [7] showed, in the case of *Bacillus stearothermophilus* growing in both the suboptimal and optimal temperature ranges for growth in glucose-limited chemostat culture, that at a dilution of $0.1\ h^{-1}$ the glucose

based biomass yield coefficient increased with increasing operating temperatures, whilst at a dilution rate of 0.2 h^{-1}, it was constant over the range of operating temperatures used. The interpretation of this behaviour was based on a restricted analysis of endogenous events.
The most plausible explanation for temperature associated depression of the biomass yield coefficient in chemostat cultures of *Bacillus* sp. NCIB 12522 is a combination of endogenous and exogenous activity, in spite of the fact that *Bacillus* sp. NCIB 12522 is relatively restricted with respect to its carbon energy substrate utilization spectrum [1].

Conclusions

In spite of the relative fastidiousness that the thermotolerant methylotrophic *Bacillus* sp. NCIB 12522 exhibits in batch culture, its response to growth at various temperatures and at two different dilution rates in chemostat culture was entirely predictable. However, the declining biomass yield coefficient within its optimum temperature range for growth at the lower dilution rate, 0.22 h^{-1}, suggests that certain anomalies exist between growth in batch and growth in chemostat culture for this bacterium. Possible extracellular product formation at higher growth temperatures and at higher growth rates and possible reutilization need further investigation.

Acknowledgement

Our thanks are due to Mrs. R. Meisser for typing the manuscript. N. Al-A. was the recipient of a Kuwait Institute for Scientific Research (KISR) research scholarship.

Received December 29, 1988

References

[1] Al-Awadhi, N., Egli, Th., Hamer, G., Wehrli, E.: Systematic Appl. Microbiol. **11** (1989), 207.
[2] Al-Awadhi, N., Egli, Th., Hamer, G.: Appl. Microbiol. Biotechnol. **29** (1988), 485.
[3] Pirt, S. J.: Principles of Microbe and Cell Cultivation. Blackwell, Oxford, GB, 1975.
[4] Hamer, G.: Acta Biotechnol. **5** (1985), 117.
[5] Mason, C. A., Hamer, G.: Dechema Monographs. **105** (1987) 285.
[6] Herbert, D.: In Continuous Culture 6, eds. A.C.R. Dean, D. C. Ellwood, C. G. T. Evans & J. Melling, SCI, London/Ellis Horwood, Chichester (1976), 1.
[7] Pennock, J., Tempest, D. W.: Arch. Microbiol. **150** (1988) 452.
[8] Mainzer, S. E., Hempfling, P.: J. Bacteriol. **126** (1976), 251.
[9] Topiwala, H., Sinclair, C. G.: Biotechnol. Bioeng. **13** (1971), 795.
[10] Matsché, N. F., Andrews, J. F.: Biotechnol. Bioeng. Symp. **4** (1973), 77.
[11] Kuhn, H. J., Cometta, S., Fiechter, A.: Eur. J. Appl. Microbiol. Biotechnol. **10** (1980), 303.

Acta Biotechnol. 10 (1990) 1, 79–84 Akademie-Verlag Berlin

The Microbiological Data Base of the Data Bank Project BD1

BRÜGGEMANN, L., SOMMER, S., FIEDLER, E., KARBAUM, K., ALBRECHT, C., HEDLICH, R., NAGEL, B.

Academy of Sciences of the G.D.R.
Institute of Biotechnology
Permoserstraße 15, Leipzig, 7050 G.D.R.

Summary

The aim of the project BD1 (Biotechnology Data Bank 1) is to improve the data processing for research in biotechnology. In BD1, data of several classes of objects and methods are compiled. A data file conception consisting of object data files, method data files and feature data files was created, which can be easily extended to further classes, and which allows the combination of different classes of features to the same class of objects. There are two main approaches to use BD1: 1. making available informations on certain biotechnological objects, 2. the identification of unknown objects by means of their feature pattern.
The microbiological data base and the possibilities of data base search are presented by example of lactic acid bacteria.

Introduction

During the last three decades, progress has been reached in computerized information processing in microbiology. Main points in this development are the application of methods of numerical taxonomy by SOKAL and SNEATH [1], the comprehensive coding of microbiological informations and the creation of microbiological data bases by ROGOSA et al. [2] and currently the creation of the Microbial Information Network Europe (Project MINE of leading microbiological organizations from six West-European countries) [3].
The present state of information processing in microbiology [4] comprises essentially three classes of methods:

- information search in data bases and expert systems,
- application of pattern recognition methods (e.g. reference search, cluster algorithms),
- application of mathematical-statistical methods of multivariate data analysis (e.g. principal component analysis).

The increasing use of microorganisms in biotechnology is an essential reason and justification for this development. Thus, the need for the exact differentiation and characterization of microorganisms has increased. At the same time, with the application of modern analytical methods to microorganisms, the possibilities for research in the field of microbiological taxonomy have become greater [5].

In this context, problems of data storage, of systematic data management including the means used to record and access these data, of statistical data analysis and other mathematical methods applied for the exploitation of information content as well as for the combination of different classes of characteristic features have to be solved.
To assist at solving of this problems, the data bank project BD1 was created [6]. Main aim of the project is to improve information processing in microbiology and biotechnology with regard to rapidity, accuracy, reliability and economy. In the first instance, problems should be solved such as the characterization and selection of microorganisms, as well as monitoring of microbial growth and product formation processes.

Basic Principles of BD1

BD1 is created as factographic data base for several classes of objects, features and methods. The respective classes of data can be combined, interrelated and expended. At present, the following main classes are included:

Main classes of objects	Main classes of methods
microorganisms	microbiological methods
biomasses and products	spectroscopic methods
chemical compounds	chromatographic methods
	methods of chemical analytics

The main classes of objects are subdivided into object classes, and the main classes of methods into feature classes. For the object classes, object data files were built up consisting of an identificator and object-related informations such as denotations, positions within systematics and biotechnologically relevant properties.
For the main classes of methods, methods data files were built up consisting of a systematic method key as identificator and method-related informations, such as testing methods, cultivation standards, conditions of measuring and analyzing. Generally, by means of the assignment of feature data files containing qualitative and quantitative features to object classes or feature classes, the object classes and feature classes can be combined conveniently.
Moreover, storage and processing both of results of current investigation and literature data are possible. Standardized abbreviations, catalogized object keys, method keys and feature keys, as well as systematic file denominations from which the data source can be seen, are used.
For BD1, there are two typical methods of operation:

1. making available informations of present (known) objects, methods and biotechnological parameters by means of identificators,
2. identification and classification of unknown objects by means of recognition of feature patterns.

For the treatment of these data base, e.g., for typical object-feature relationships, a relational data model is used within the data bank operation system dBASE III. Using this operation system, data input, data storage and data management as well as the search for logically linkable terms are performed including table printing operations.
A further data processing with regard to data reduction and data analysis by preferentially used mathematical-statistical methods is carried through by means of PASCAL programs [7] and statistics software processing the dBASE feature data files. Moreover, BD1 is a basis for the application of mathematical methods of pattern recognition for the identification and classification of objects.

Structure of Microbiological Data

The main class of objects "microorganisms" was subdivided into the following object classes:

11 yeasts
12 fungi (except yeasts)
13 algae
14 gram-negative aerobic rods and cocci (Section 4; BERGEY 1984 [8])
15 gram-positive cocci (Section 12)
16 endospore-forming gram-positive rods and cocci (Section 13)
17 regular, non-sporing gram-positive rods (Section 14).

These object classes can be combined with microbiological, chromatographic and spectroscopic methods. In the class of microbiological methods, the feature classes 11 (physiological features) and 12 (morphological features) were established. The identificator for the microbiological objects consists of three fields illustrated as follows:

Aggregation level	STRAIN	SPECIES	GENUS
STRAIN	IBT B 1234	0123	1301
SPECIES	IBT B 0000	0123	1301
GENUS	IBT B 0000	0000	1301
SECTION	IBT B 0000	0000	1300

The microbiological features were catalogized as feature catalogue using the following systematic key:

1. digit ... letter for group (e.g. P = Physiological-biochemical characteristics)
2.—3. digit ... current number of subgroup (e.g. 0 = Acid from)
4.—5. digit ... current number of feature (e.g. 01 = D-Glucose)

The feature catalogue can be expanded. It contains only those features which are really used for identification. Both the physiological-biochemical and the morphological feature distinctions were nominally scaled encoded by the categories '+' (positive reaction), '—' (negative reaction), 'V' (variable) and '?' (unknown). Moreover, the category '+' was subdivided by several graduations. The identificator for the microbiological methods contains digits which code the object classes and feature classes as well as a mark for the data source, a current number of method (e.g., physiological features from BARNETT [9]) and referring to the feature catalogue, a feature code (e.g.,

Tab. 1. Object data file (lactic acid bacteria)

	BD1-OBJECT DATA FILE D01700		
	STRAIN IBT B 1030 SPECIES 0025 GENUS 1401 SECTION 14		
	CD 07/88 TS		
NAME	*Lactobacillus plantarum*		
NUMBERS	IBT, 30		
DESCRIPT	ALBRECHT, CH.; G-3, 1987		
HISTORY	1985, IBT		
ORIGIN	milk products		
UTILISAT	glc	PROD lactic a.	
CULTIVATION CONDITIONS:	TEMP 15 37 50 pH 04,0 06,8	MED 001	
MAINTENANCE:	002,14 d,4 C; lyophil.		OECO microaeroph.
SPEC			

P0200 means, that the method-related data refer to the subgroup "Acid from"). Specific informations of the method, which provides the measuring values, can be represented format-free in two fields. By means of object identification and method identification, the feature data files can be related to the respective object data file or method data file. Cultivation methods which were applied, biotechnologically important properties and several abbreviations were also catalogized and standardized. The structure of microbiological data of lactic acid bacteria is given in Tab. 1—4, as an example.

Tab. 2. Feature data file (lactic acid bacteria)

BD1-FEATURE DATA FILE DF1711
STRAIN IBT B 1030 SPECIES 0025 GENUS 1401 SECTION 14 METHKEY 1002
FEATURES:
P0201 + P0202 + P0203 + P0205 + P0206 + P0207 + P0208 + P0209 + P0210 − P0211 + P0212 − P0213 − P0214 + P0215 + P0216 + P0217 + P0218 − P0220 − P0221 + P0222 V P0223 + P0228 − P0232 − P0502 − P0504 − P0505 − P0508 + P0511 + P0522 + P0524 + P0526 + P0528 + P0529 − P0602 − P0606 − P0901 + P1002 − P1003 − P1004 + P1101 − P1102 + P1103 − P1201 +

Tab. 3. Method data file (microbiological methods)

BD1-METHOD DATA FILE DM0010
METHKEY: 17111002 F-CODE: P0200 METHOD: SHARPE, M.E.; 1962
MODIFY:

Tab. 4. Feature Catalogue

Physiological-biochemical characteristics	P0000
Acid from	P0200
D-Glucose	P0201
D-Galactose	P0202
Maltose	P0203
Me-α-D-glucoside	P0204
Sucrose	P0205
α, α-Trehalose	P0206
Melibiose	P0207
Lactose	P0208
Cellobiose	P0209
Melezitose	P0210
,,	,,
,,	,,
,,	,,

Search in the Microbiological Data Base

In the following, the typical search possibilities are demonstrated:

Search for Product Formation Properties

To initiate industrial fermentations, authentic preparations of a known organism are used as starter cultures, e.g. for production of cheese. Lactic acid bacteria are required, they should be isolated from cheese, should be thermotolerant (growth at 45 °C) and halotolerant (NaCl concentration $< 6\%$). The keyword "ORIGIN" is used to search for "cheese" in the object data file. Then, the strains under consideration are linked with the properties thermotolerance (P0511) and halotolerance (P0524 and P0526) from the

feature data file, and the computer lists all strains meeting the requirements mentioned above.
As a result, two strains meet the condition "cheese", however, only one of them is thermotolerant and halotolerant. The searching run algorithm is demonstrated in Tab. 5.

Tab. 5. Search run for product formation properties

```
• USE D01700
• SET FORMAT TO DF1711M4
• LOCATE FOR ORIGIN = 'cheese'
Record =        10
• edit
                STRAIN:IBT B 1010 SPECIES:0009   GENUS:1401
• CONT
Record =        11
• edit
                STRAIN:IBT B 1011 SPECIES:0028   GENUS:1401
• CONT
End of LOCATE scope
• USE DF1711
• SET FORMAT TO DF1711M4
• LOCATE FOR (STRAIN='IBT B 1010' .OR.STRAIN='IBT B 1011').AND.P0511='+
'.AND.P0524='+'.AND.P0526='+'
Record =        4
• edit
                STRAIN:IBT B 1010 SPECIES:0009   GENUS:1401
• CONT
End of LOCATE scope
```

Identification of Unknown Objects

At the isolation of microorganisms from natural biotops, several strains must be assigned taxonomically to the species level. For this purpose, the results of diagnostic tests are compared with the features of known strains from cultur collections.
To identify the species, in the following example the condition "milkproducts" from the ORIGIN field of the object data file is combined with the feature pattern
P0211 = '+', P0511 = '+', P1004 = '+', P1101 = '–'
of the feature data file.
Two strains meet completely the conditions identified above. Therefore, they can be considered as identical with the isolated strain. The result of the search run is presented in lucid form by means of a listing generator.
Feature patterns which are required for identification can also be calculated by means of mathematical logic and fuzzy technique as minimal feature sets [7]. The search run algorithm can be seen from Tab. 6–7.

Tab. 6. Search run for identification

```
• SELECT 2
• USE DF1711
• SELECT 1
• USE D01700
• JOIN WITH DF1711 TO DF1711T5 FOR SPECIES=DF1711→SPECIES.AND.ORIGIN=
'milk products' .AND.DF1711→P0208='+'.AND.DF1711→P0211='+'.AND.DF1711
→P0511='+'.AND.DF1711→P1004='+'.AND.DF1711→P1101='–' FIELDS
STRAIN, SPECIES, ORIGIN, P0208, P0211, P0511, P1004, P1101
      2 records joined
```

Tab. 7. Identification search result

Page No. 1
10/28/88

IDENTIFICATION OF UNKNOWN ISOLATES

		ACID FROM		GROWTH AT	DL-LACTIC A.	CO 2 FROM
STRAIN	SPECIES ORIGIN	LACTOSE	RAFFINOSE	45 C		GLUCOSE
IBT B 1001 0025	milk products	+	+	+	+	−
IBT B 1030 0025	milk products	+	+	+	+	−

Received November 21, 1988

References

[1] Sokal, R. R., Sneath, P. H. A.: Principles of Numerical Taxonomy, Freeman, San Francisco, 1963.

[2] Rogosa, M., Krichevsky, M. I., Colwell, R. R.: Coding Microbial Data for Computers, Springer-Verlag, Berlin 1987.

[3] Gams, W., Hennebert, G. L., Stalpers, J. A., Jansens, D., Schipper, M. A. A., Smith, I., Yarrow, D., Hawksworth, D. L.: J. Gen. Microbiol. **134** (1988), 1667.

[4] Abstracts of 2nd Conference on Taxonomy and Automated Identification of Bacteria, Prague, 1987.

[5] Nagel, B., Schmidt, A., Flemming, Ch., Geyer, W., Mai W., Brüggemann, L.: Wiss. Z. Karl-Marx-Univ. Leipzig, Math.-Naturwiss. Reihe **38** (1989), 327.

[6] Brüggemann, L., Fiedler, E., Karbaum, K.: Poster, Taxonomy Conference Prague, 1987.

[7] Brüggemann, L., Sommer, S., Flemming, I., Nagel, B.: J. System. Appl. Microbiology, submitted.

[8] Krieg, N. R. (Editor Volume 1), Holt, I. G. (Editor in Chief): Bergey's Manual of Systematic Bacteriology, Williams & Wilkins, Baltimore/London, 1984.

[9] Barnett, J. A., Payne, R. W., Yarrow, D.: Yeasts-characteristics and identification, Cambridge University Press, Cambridge, London, New York, 1983.

Acta Biotechnol. 10 (1990) 1, 85—92 Akademie-Verlag Berlin

Nonionic Tensides Modify Papain Structure and Proteolytic Activity

SZÖGYI, M.[1], CSERHÁTI, T.[2]

[1] Institute of Biophysics, Semmelweis Medical University
Budapest, P.O.B. 263. H-1444 Hungary
[2] Central Research Institute for Chemistry
Hungarian Academy of Sciences
H-1525 Budapest, P. O. Box 17, Hungary

Summary

The effect of 33 polyethoxylated nonionic tensides with various hydrophobic moieties was studied on the proteolytic activity and phase transition behaviour of papain. Tensides with longer ethyleneoxide chain markedly increased the phase transition temperatures of papain indicating possible hydrogen bond formation between the hydrophilic ethyleneoxide chain and the polar substructures of papain. The character of lipophilic moiety of tensides influences also each physicochemical parameter of papain suggesting the existence of hydrophobic interactions too. The significant positive linear correlations between the activity increasing and structure stabilizing effects of tensides make probably that the structure modification of papain accounts for its enchanced proteolytic activity.

Introduction

Nonionic tensides are widely used for many purposes, as in pesticide and pharmaceutical formulations. The concomitant toxic phenomena are also varied. Although they are easily biodegredable they nevertheless decrease the performance of activated sludge systems [1, 2], damage pine trees by increasing the uptake of sodium ions [3, 4, 5], and by dissolving the surface wax layer of pine needles [6]. Tensides exert an inhibitory effect on the growth of *Spirodela polyrrhiza* (L) Schleiden [7], they are lethal to Medaka Oryzias Latipes [8] inhibit the potassium uptake of wheat seedling roots [9], significantly enhance absorption rate constant for highly lipophilic 4-alkylanilines in rat colon [10], modify the dissolution rate and disintegration of granules and tablets [11] they are cytolytic [12] etc.

The biological activity of nonionic tensides is based on their interaction with membrane phospholipids [8, 13], however it has been reported, that they modify protein conformation [14], disrupt phospholipid-protein complexes [15] and enhance enzymatic activity [16].

The objectives of our investigation were to study the interaction of some nonionic tensides with the proteolytic enzyme papain and to correlate the structure modifying effect of tensides with their influence on the proteolytic activity.

Material and Methods

Papain was purchased from SIGMA CHEMICAL Co. (USA) and used without further purification. The nonionic tensides were polyethoxylated derivatives as Tween 20, 40, 60 and 80 (ATLAS, FRG), Genapol O 20, 50, 80, 150 and 200 (HOECHST FRG), Arkopal R 40, 50, 60, 80, 90, 100, 110, 130, 150, 230 and 300 (HOECHST, FRG), Sapogenate T 40, 60, 80, 110, 130, 180, 300 and 500 (HOECHST, FRG), and Myrj 49, 51, 52, 53 and 59 (ATLAS, FRG). Their chemical structures are shown in Tab. 1.

Tab. 1. Chemical structure of nonionic tensides
$Q - O(C_2H_4O)_{n_e}-H$

Common name	General structure	
	Q	n_e (average)
Tween 20	anhydrosorbit monolaurate	20
Tween 40	anhydrosorbit monopalmitate	20
Tween 60	anhydrosorbit monostearate	20
Tween 80	anhydrosorbit monooleate	20
Genapol O 20	oleylalcohol	2
Genapol O 50	oleylalcohol	5
Genapol O 80	oleylalcohol	8
Genapol O 150	oleylalcohol	15
Genapol O 200	oleylalcohol	20
Arkopal N 40	nonyl-phenol	4
Arkopal N 50	nonyl-phenol	5
Arkopal N 60	nonyl-phenol	6
Arkopal N 80	nonyl-phenol	8
Arkopal N 90	nonyl-phenol	9
Arkopal N 100	nonyl-phenol	10
Arkopal N 110	nonyl-phenol	11
Arkopal N 130	nonyl-phenol	13
Arkopal R 150	nonyl-phenol	15
Arkopal N 230	nonyl-phenol	23
Arkopal N 300	nonyl-phenol	30
Sapogenate T 40	tributhyl-phenol	4
Sapogenate T 60	tributhyl-phenol	6
Sapogenate T 80	tributhyl-phenol	8
Sapogenate T 110	tributhyl-phenol	11
Sapogenate T 130	tributhyl-phenol	13
Sapogenate T 180	tributhyl-phenol	18
Sapogenate T 300	tributhyl-phenol	30
Sapogenate T 500	tributhyl-phenol	50
Myrj 49	stearic acid	20
Myrj 51	stearic acid	30
Myrj 52	stearic acid	40
Myrj 53	stearic acid	50
Myrj 59	stearic acid	100

For differential scanning calorimetry (DSC) the samples were prepared by mixing the papain and tensides with twice distilled water for 30 min with a vortex mixer. The papain: tenside molar ratio was 1:5. The measurements were carried out in a DuPont 910 DSC cell at a heating rate 5 °C/min and in the sensitivity range of 0.1—0.2 mW/cm. The two transition temperatures (T_{p_1}, T_{p_2}) of papain were determined.

The effect of tensides on the proteolytic activity of papain was determined on caseine substrate. Tensides, papain and caseine were dissolved in 0.16 M phosphate buffer (pH = 7.0), the mixture was incubated for an hour at 37 °C than the activity was determined by the BIURET reaction [7]. The more sensitive FOLIN reagent reacted with the tensides therefore its application was impossible in our case. The tenside-papain molar ratio was 135:1. In case of Arkopal N90 the tenside-papain molar ratio varied from 10:1 to 220:1. Each experiment (DSC and determination of proteolytic activity) was run in quadruplicate.

To elucidate the relationship between the influence of tensides on the proteolytic activity ($A\%$) and on the phase transition temperatures (T_{p_1} and T_{p_2}) of papain linear correlations were calculated between the variables mentioned above:

$$A\% = a_1 + b_1 T_{p_1} \quad (1)$$

$$A\% = a_2 + b_2 T_{p_2}. \quad (2)$$

To asses the impact of various substructures of tenside molecules on the proteolytic activity and on the phase transition temperatures of papain stepwise regression analysis was applied [18]. The calculation were carried out three times, the proteolytic activity and the two phase transition temperatures of papain beeing the dependent variables. The independent variables were in each case the type of hydrophobic moiety of tensides (the presence of the bulky anhydrosorbate, nonylphenyl and tributhylphenyl groups, the number of carbon atoms in the alkyl chain, the alcoholic or acidic character of the hydrophobic moiety and the presence of double bond in the alkyl chain) and the linear, logarithmic and quadratic forms of the average number of ethyleneoxide groups per molecule. The number of accepted variables was not limited, the partial F value of independent variables was set to $F = 2$.

Results and Discussion

Papain showed two transition temperatures in DSC experiments (Fig. 1). This result indicates that the structure modification of papain takes place in two separate steps, however, our data are not suitable to determine the exact structural changes of papain corresponding to the phase transition temperatures observed. Tensides increase the transition

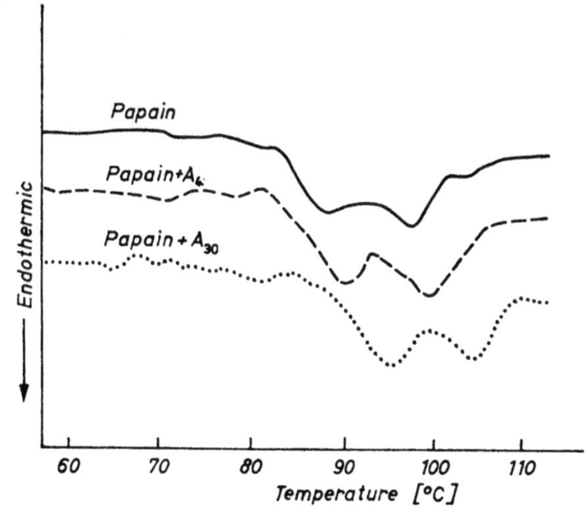

Fig. 1. Effect of two nonylphenyl ethyleneoxide polymers ($n_e = 4$ and 30) on the phase transition behaviour of papain

temperatures of papain that is the tenside-papain interaction stabilizes papain structure. Tensides not only stabilize papain structure but also considerably enhance the proteolytic activity (Fig. 2). The effect non-linearly increases with increasing tenside: papain molar ratio, no saturation was achieved even at 220:1 tenside: papain molar ratio. This observation suggests, that the tensides interact with many sites of the papain peptide chain, these interactions probably are of different interactive strength and of different character resulting in the monotonous but non-linear increase of papain activity.

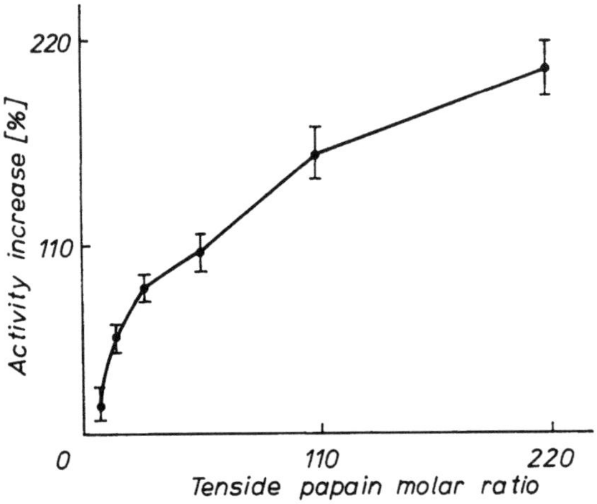

Fig. 2. Effect of various quantities of nonylphenyl nonyl-glycolate on the proteolytic activity of papain

Our results are compiled in Tab. 2. Except Sapogenate T 40, Mirj 49 and Tween 80 each tenside markedly enhanced the proteolytic activity of papain, the activity increased with increasing length of the hydrophilic ethyleneoxide chain of tenside. This phenomenon suggests, that hydrophilic interaction (probably hydrogen bonds) may exist between the ethyleneoxide chains of tensides and the polar parts of papain molecule. However, the tensides with various hydrophobic moieties and identical ethyleneoxide chains (Tween 20, 40, 60 and 80, Genapol O 200 and Myrj 49) exerted different impact on the proteolytic activity, that means, that not only hydrophilic but also hydrophobic interactions may exist between papain and tensides. The effect of tensides on both phase transition temperatures supports our previous considerations, the tensides with longer ethyleneoxide chain increased more markedly the phase transition temperatures suggesting stronger interactions, however, tensides with identical ethyleneoxide chain and various hydrophobic moieties differently influenced the phase transition behaviour of papain.

It can be concluded from our data that minimally two types of interactions prevail between nonionic tensides and papain: hydrophilic interaction between the ethyleneoxide chain of tensides and the more polar parts of papain, hydrophobic interaction between the hydrophobic core of papain and the lipophilic moiety of tensides. Both interactions influence the proteolytic activity and phase transition characteristics of papain.

The parameters of equ. 1. and 2. are compiled in Tab. 3. In both cases significant linear correlations (significance level over 95%) were found between the effect of tensides on the proteolytic activity and on the phase transition temperatures of papain. The tensides which increase more strongly the proteolytic activity also stabilize more strongly

Tab. 2. Effect of some nonionic tensides on the proteolytic activity and phase transition temperatures of papain
Papain $T_{p_1} = 88.0 \pm 1.1\,°C$, $T_{p_2} = 96.0 \pm 2.5\,°C$

Tenside		Proteolytic acitity %		T_{p_1} °C		T_{p_2} °C	
		$\bar{x}$	s	$\bar{x}$	s	$\bar{x}$	s
Tween	20	119.3	1.0	91.8	0.4	101.3	0.4
	40	159.5	5.5	95.3	1.1	102.8	0.3
	60	129.9	3.7	94.5	0.7	101.8	0.4
	80	86.6	6.8	92.3	1.1	101.0	0.1
Genapol O	20	121.1	5.7	89.0	1.4	98.0	0.6
	50	124.5	4.2	91.5	0.7	101.2	0.5
	80	142.6	5.1	92.7	0.4	102.3	0.4
	150	127.5	3.4	93.5	0.8	102.7	0.3
	200	107.3	4.0	93.8	1.1	102.8	1.3
Arkopal N	40	120.2	2.7	90.3	1.8	99.0	1.3
	50	128.5	6.2	89.8	0.4	99.7	0.4
	60	128.8	4.4	92.3	0.3	101.2	0.4
	80	133.2	5.1	93.2	0.2	102.3	0.3
	90	136.3	4.1	93.0	1.5	102.3	0.4
	100	135.8	3.4	92.2	0.3	101.7	0.2
	110	151.4	3.1	93.3	0.3	102.5	0.6
	130	169.5	6.7	93.6	0.6	102.6	0.8
	150	158.8	4.4	94.5	2.1	103.0	2.8
	230	188.9	10.1	93.7	0.3	103.5	2.0
	300	193.0	2.2	95.5	0.7	104.8	0.3
Sapogenate T	40	92.5	2.1	89.3	0.4	98.1	0.6
	60	101.2	6.1	90.0	0.6	98.5	0.4
	80	112.0	2.3	90.8	1.1	100.5	0.5
	110	138.8	4.1	92.5	0.4	102.2	0.2
	130	157.8	6.0	92.2	0.3	102.3	0.4
	180	170.1	5.6	91.1	0.6	101.3	0.4
	300	197.3	3.6	92.0	1.1	102.2	0.3
	500	231.5	3.4	91.3	0.4	101.7	0.4
Myrj	49	63.2	3.7	91.0	0.5	101.5	0.6
	51	139.9	5.0	92.5	0.4	102.5	0.7
	52	162.4	13.9	93.5	0.6	103.0	1.1
	53	137.7	2.5	96.5	0.4	105.0	1.0

Tab. 3. Linear correlations between the effect of tensides on the proteolytic activity and on the phase transitions parameters of papain $r_{95\%} = 0.3494$

	Eq. 1.	Eq. 2.
a	−502.0	−113.4
b	6.94	10.35
s_b	3.26	3.33
$r_{calc.}$	0.3621	0.4935

the papain structure. These significant correlations suggest that the structure stabilizing effect of tensides may account for the increased proteolytic activity, that is in our case the structure modification results in modified function. The results of stepwise regression analysis are compiled in Tab. 4, 5 and 6. In each case the equations selected by the stepwise regression analysis fit well to the experimental data, the significance level was always over 99.9% ($F_{calc} > F_{99,9\%}$). The change of the independent variables accounts for about 59—77% of the change of various dependent variables (see r^2 values). The path coefficients ($b'\%$ values) prove that the hydrophilic interactions (probably hydrogen bonds) generally have a higher impact on the papain-tenside interactions than the hydrophobic bonding forces (lower path coefficients for the lipophilic anhydrosorbate, tributhylphenyl and nonylphenyl moieties). Besides the number of ethyleneoxide groups other molecular substructures also influence significantly the various physico-chemical parameters of papain. This result can be tentatively explained by the assumption that tensides with various hydrophobic moieties preferentially bond to the different parts of papain. In other word the distribution of bonded tenside molecules along the papain molecule may be different for tensides with various hydrophobic part resulting in various influencing of physico-chemical parameters of papain. Our data indicate that tensides with anhydrosorbate moiety bond more probably to the papain region respon-

Tab. 4. Effect of tenside substructures on the proteolytic activity of papain. Results of stepwise regression analysis

$A\% = a + b_1x_1 + b_2x_2 + b_3x_3 + b_4x_4$

x_1 = number of anhydrosorbate ring in the tenside
x_2 = number of ester bond in the hydrophobic moiety
x_3 = number of double bond in the alkyl chain
x_4 = number of ethyleneoxide group per molecule

$a = 115.1$ $F_{calc.} = 16.41$ $F_{99.9\%} = 6.33$ $r^2 = 0.7089$ $t_{95\%} = 2.05$

	x_1	x_2	x_3	x_4
b	39.76	−76.36	−19.92	2.49
s_b	15.11	13.15	9.27	0.35
$b'\%$	15.5	39.1	9.2	36.2
$t_{calc.}$	2.63	5.80	2.15	7.04

Tab. 5. Effect of tenside substructures on the first phase transition temperature of papain. Results of stepwise regression analysis

$T_{p_1} = a + b_1x_1 + b_2x_2$

x_1 = number of tributhylphenyl moiety
x_2 = logarithm of the number of ethyleneoxide group per molecule

$a = 89.3$ $F_{calc.} = 20.93$ $F_{99.9\%} = 8.85$ $r^2 = 0.5906$ $t_{99\%} = 2.76$

Parameter	Independent variable	
	x_1	x_2
b	−1.71	3.39
s_b	0.48	0.63
$b'\%$	39.60	60.40
$t_{calc.}$	3.54	5.38

Tab. 6. Effect of tenside substructures on the second phase transition temperature of papain. Results of stepwise regression analysis

$T_{p_2} = a + b_1x_1 + b_2x_2 + b_3x_3$

x_1 = number of ether bond in the hydrophobic moiety
x_2 = number of nonylphenyl moiety
x_3 = logarithm of the number of ethyleneoxide groups per molecule

$a = 95.7$ $F_{calc.} = 31.94$ $F_{99.9\%} = 7.12$ $r^2 = 0.7735$ $t_{99\%} = 2.76$

Parameter	Independent variable		
	x_1	x_2	x_3
a	1.59	1.61	4.72
s_b	0.46	0.34	0.49
$b'\%$	19.7	26.20	54.00
$t_{calc.}$	3.44	4.67	9.64

sible for the proteolytic activity. The tributhylphenyl- and nonylphenyl-derivatives influence more strongly the first and second phase transition temperatures suggesting that they are preferentially bound to papain parts responsible for these phase transitions.

Summarizing our data we established that nonionic tensides significantly enhance the proteolytic activity of papain and stabilize its structure, this structure stabilization may account for the increased proteolytic acitivity. The papain-tenside interaction is mainly of hydrophilic character, however the lipophilic moiety of tensides influences also the strength of interaction.

Acknowledgements

The author wish to thank Prof. Rontó for valuable discussion and Mrs. Bányay, Mrs. Matus, Mrs. Bogdányi for technical assistance.

Received December 13, 1988

References

[1] Hashinaga, T., Dazai, M., Uehara, M.: Rep. of Ferment. Res. Inst. Japan **62** (1984), 55.
[2] Gilbert, P. A., Pettigrew, R.: Int. J. Cosmetic Sci. **6** (1984), 149.
[3] Gellini, R., Pantani, F., Rossoni, P. G., Bussotti, F., Barbolani, E., Rinallo, C.: Eur. J. For. Pathol. **13** (1983), 296.
[4] Truman, R., Lambert, M. J.: Aust. J. Plant. Physiol. **5** (1978), 337.
[5] Dowden, H. G. M., Lambert, M. J., Truman, T.: Aust. J. Plant. Physiol. **5** (1978), 387.
[6] Gellini, R., Pantani, F., Grossoni, P., Bussotti, F., Barbolani, E., Rinallo, C.: J. For. Pathol. **15** (1985), 145.
[7] Buczek, J.: Acta Soc. Bot. Pol. **53** (1984), 551.
[8] Kikuchi, M., Wakabayashi, M.: Bull. Jpn. Soc. Sci. Fish. **50** (1984), 1235.
[9] Bujtás, K., Cserháti, T., Szejtli, J.: J. Incl. Phenom. **5** (1987), 421.
[10] Plá-Delfina, J. M., Pérez Buendia, M. O., Casabo, V. G., Perisribera, J. E., Sanchez-Moyano, E., Martin-Villodre, A.: Int. J. Pharm. **37** (1987), 49.

[11] WANRAND, L. S. C., HENG, P. W. S.: Pharm. Acta. Helv. **62** (1987), 168.
[12] RÖDERER, G.: Arch. Environ. Contam. Toxicol. **16** (1987), 291.
[13] HELENIUS, A., SIMONS, K.: Biochim. Biophys. Acta **415** (1975), 29.
[14] MAMMI, S., MAMMI, N. J., TOFFANI, T., PEGGION, E., MORODER, L., WÜNSCH, E.: Biopolymers **26** (1987), 41.
[15] HOLMES, E. H., HAKOMORI, S. I.: J. Biochim. **101** (1987), 1095.
[16] NAKAHARA, H., OKADA, S., OHMORI, H., MASUI, M.: Chem. Pharm. Bull. **32** (1984), 3803.
[17] KAKÁC, B., VEGDEELEK, Z. I. — In: Handbuch der Kalorimetrie Band III. VEB Gustav Fischer Verlag, Jena, 1966, 235.
[18] MAYER, H. — In: Moderne Regressionsanalyse Salle, Sauerländer Frankfurt am Main, 1982, 135.

Acta Biotechnol. 10 (1990) 1, 93–98 Akademie-Verlag Berlin

Isolation of Auxotrophic Mutants and Development of Conjugation System in KISR Methylotrophs

KADRI, M.[1], AL-SALAMEEN, F.[1], HADITIRTO, S.[1], ENGEL, J.[2]

[1] Kuwait Institute for Scientific Research, P. O. Box 24885, 13109 Safat Kuwait.
[2] Academy of Sciences of the G.D.R.
Institute of Biotechnology
Permoserstr. 15, Leipzig, 7050 G.D.R.

Summary

Isolation of auxotrophic mutants and conjugal transfer of plasmids are successfully demonstrated in four obligate methylotrophic strains isolated at Kuwait Institute for Scientific Research (KISR). Out of 11 auxotrophic, mutants, three required methionine and two leucine. Complete transfer of plasmid R68.45 from *Pseudomonas* to KISR methylotrophs and vice versa have been achieved with a low transfer rate.

Introduction

Four novel methanol-utilizing bacterial strains, namely KISRI-5, KISRI-6.1, KISRI-512 and KISRI-5112 were isolated from Kuwaiti soils and developed for the production of single cell protein (SCP) during the earlier research work at KISR. Detailed morphological and biochemical characterization studies for the four strains have been carried out and they were found to belong to the genus *Methylophilus* [1]. These strains were deposited for patent purposes in the National Collection of Industrial and Marine Bacteria Ltd. (NCIMB) and have been designated with the following numbers: NCIMB-12135 (KISRI-5), NCIMB-12136 (KISRI-6.1), NCIMB-12137 (KISRI-512) and NCIMB-12138 (KISRI-5112) [2]. These findings gave an impetus for further studies on genetic characterization of these strains with the aim of isolating mutants with improved characteristics and the development of conjugation system.

Isolation of mutants from methylotrophs using ultraviolet (UV) irradiation and chemical mutagen N-methyl-N-nitro-nitrosoguanidine (MNNG) have been reported earlier [3, 4, 5]. The transfer of plasmid DNA in methylotrophs by conjugation have been reported by ENGEL [6], HOLUBOVA [7] and WARNER [8]. This communication reports the preliminary genetic analysis of the four novel obligate KISR methylotrophs by isolating the auxotrophic mutants and testing their recipients' ability of foreign DNA by conjugation.

Materials and Methods

Bacterial Strains and Growth Condition

The strains of KISR methylotrophs, other methylotrophic bacteria and *Pseudomonas* used in this study are listed in Tab. 1. All methylotrophic strains were grown at 37 °C on 3M basal medium and on solid medium containing 2M basal medium with 1.6% agar, supplemented with 1% metha-

Tab. 1. Methylotrophic and other strains used in this study

Strain	Source
KISRI-5 (NCIMB-12135)	this laboratory
KISRI-512 (NCIMB-12137)	this laboratory
KISRI-5112 (NCIMB-12138)	this laboratory
KISRI-6.1 (NCIMB-12136)	this laboratory
Methylophilws methylotrophs (NCIMB-10515)	National Collections of Industrial & Marine Bacteria Ltd., Aberdeen
Ps. aeruginosa PAO 2001 (R68.45) arg H32, str-39, Cml-2 cb, km, tc	Institute of Biotechnology Leipzig, G.D.R
Ps. aeruginosa ML 4262 (his), trp, ilv, met, rif	Institute of Biotechnology Leipzig, G.D.R.

nol (v/v) as a source of carbon and energy [1]. Other *Pseudomonas* strains were grown on LB medium [9] and for auxotrophic mutants, a complete medium (2M + 1% methanol + 0.5% yeast extract). For the identification of auxotrophs 2M + methanol agar plates containing L-amino acids (50 μg/ml), D-amino acids (100 μg/ml) and Base (20 μg/ml) were used.

Determination of Strains Resistant to Mutagens

UV and MNNG were used for the mutagenicity assay experiments. UV survival curves were determined by growing KISR methylotrophs to exponential phase and by diluting with a phosphate buffer to get the total cell count of 5×10^7 cells/ml. Cells were exposed to a UV lamp with an emission of 254 nm at a distance of 50 cm, for different time periods (30—120 s). The exposed cells were diluted in one to ten steps, and from each dilution, 0.1-ml aliquots were spread onto 2M-methanol plates for the viable count determination.
Similarly, cells were treated with different concentrations of MNNG (50—400 μg/ml) for 30 min on a shaker at 37 °C. The treated cells were washed with a phosphate buffer (pH-7) to eliminate MNNG. The viable cell count was determined by spreading 0.1-ml aliquots from serial dilutions onto 2M-methanol plates, which were incubated at 37 °C for 24 h.

Isolation of Auxotrophic Mutants

The method described by Tsygankov and Kazakova [4] was used with slight modification. Cells of KISR methylotrophs grown at their exponential phase were exposed to MNNG in a final concentration of 200 μg/ml for one hour at 37 °C, with continuous shaking on an incubator-shaker. MNNG was removed from the mutagenized cells by washing thrice with phosphate buffer (pH-7).

Tab. 2. Combination of amino acids and bases for characterizing auxotrophs*

Plate No.	1	2	3	4	5
6	adenosine	guanosine	cysteine	methionine	thiamine
7	histidine	leucine	isolleucine	lysine	valine
8	phenylalanine	tyrosine	tryptophan	threonine	proline
9	glutamine	asparagine	uracil	aspartic acid	arginine
10	thymine	serine	glutamic acid	DAP	glycine

* source of table Davis [10]

Serial dilutions were prepared and 0.1-ml aliquots were spread onto 2M+methanol+yeast extract agar plates. Colonies obtained after 2–3 days of incubation at 37 °C were replica plated onto 2M-methanol agar plates. Colonies that grow on the master plate but failed to grow on 2M-methanol plates after 2 days of incubation at 37 °C were considered as potential auxotrophic mutants. The mutants were further tested for their specific amino acid requirement. Mutant cells grown in 3M-methanol yeast extract medium were washed with phosphate buffer, and from suitable dilutions, 0.1-ml aliquots were spread onto 10 different plates containing various combinations of amino acids or bases or both, as shown in Tab. 2. 2M-methanol plates were used as negative control and 2M-methanol-yeast extract plates were used as positive control. The specific growth requirements for each mutant were determined by observing growth after 2 days of incubation at 37 °C, on a plate containing a combination of amino acids.

Conjugation Experiments

Transfer of R68.45 from Ps. aeruginosa to KISR Methylotrophs

For the conjugational transfer of R68.45 from *Ps. aeruginosa* PAO 2001 to KISR methylotrophs, tetracycline resistance was used as selective marker. Donor and recipient cells grown to their exponential phase were mixed in a proportion of 1:4. From the mixture, 0.2 ml was placed onto the conjugation plate (2M+methanol+0.1% glucose +25 µg/ml arginine). Controls were taken by spreading 0.4-ml of donor cells and 0.16-ml of recipient cells onto two separate plates of the same media. After 24 h of incubation at 37 °C, cells from all the three plates were scraped and dilutions were prepared. Appropriate dilutions of the donor-recipient mixture were spread onto the selection plates (2M-methanol+tetracycline 30 µg/ml) and onto nutrient agar containing tetracycline (30 µg/ml) and 2M-methanol agar plates, to determine the total cell counts for the donor and recipient, respectively. Transconjugants were selected from the selection plates, after 3 days of incubation at 37 °C and were further confirmed by their growth on same plates. The frequentcy of transfer rate to the recipient cells was calculated.

Transfer of R68.45 from KISR Methylotrophs to Ps. aeruginosa ML 4262

Transconjugants of KISR methylotrophs containing the plasmid R68.45 were used as donor, and *Ps. aeruginosa* ML 4262 rf^r was used as recipient. The tetracycline resistance marker was chosen as the selective marker. Mating experiments were carried out as described above, except 2M+methanol+tetracycline (30 µg/ml) + rifampicin (50 µg/ml) were used as selection plates. Controls using donors and recipients separately were run under the same conditions.

Results and Discussion

Mutagens Survival Curves

The determination of optimal conditions for mutagenesis is a prerequisite for the successful isolation of mutants. Dosis/survival curves for both UV and MNNG were constructed. All four KISR methylotrophic strains show similar survival rates with MNNG (Fig. 1). The inactivation of strains KISRI-5 and KISRI-6.1 is slightly higher than that of strains KISRI-512 and KISRI-5112. The results of irradiating four KISR methylotrophic strains with UV light for various lengths of time are shown in Fig. 2. The viable count of strains KISRI-5 and KISRI-6.1 reduced to approximately 10% of the original value within 120 s of irradiation, while strains KISRI-512 and strains KISRI-5112 reduced 40% within the same period. However, all strains show similar survival pattern upto 60 s. For mutagenicity assays, we selected MNNG over UV due to the higher mutagenic activity of the former.

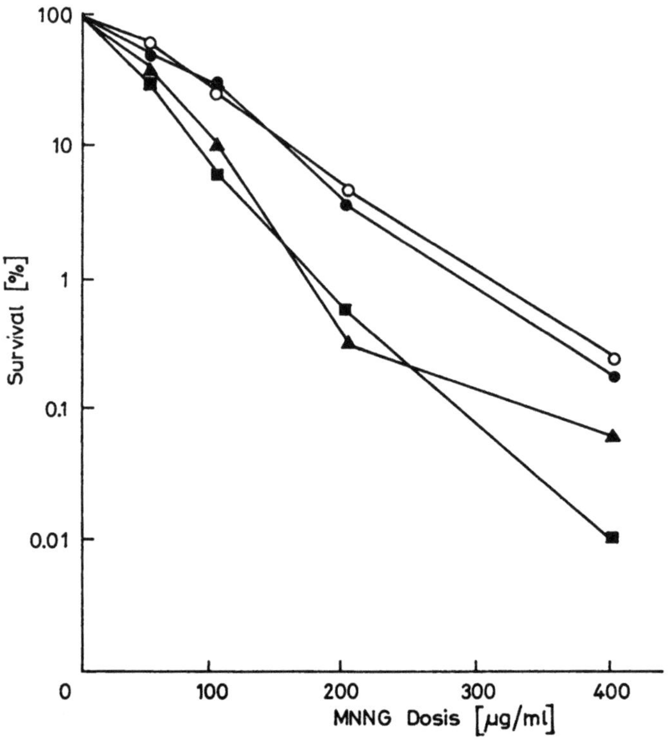

Fig. 1. MNNG doses/survival curves for strains KISRI-5 (▲), KISRI-512 (●) KISRI-5112 (○) and KISRI-6.1 (■)

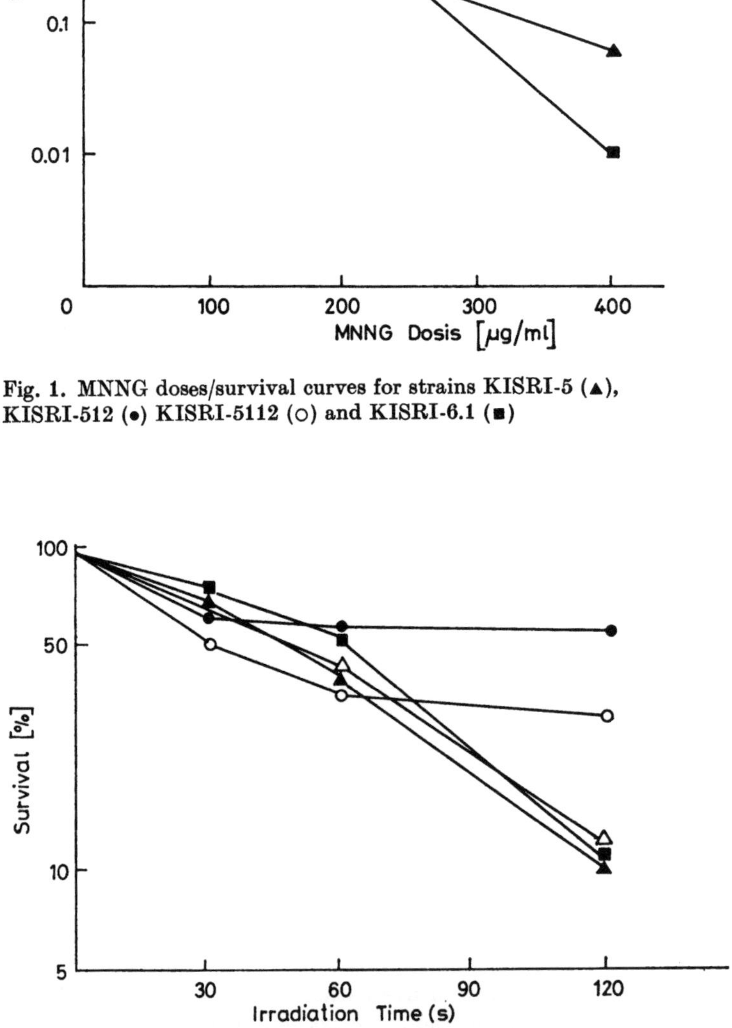

Fig. 2. UV doses/survival curves for strains KISRI-5 (▲), KISRI-512 (●), KISRI-5112 (○), and KISRI-6.1 (■)

Isolation of Auxotrophic Mutants

The experimental work was concentrated on KISRI-5 strain due to its low polysaccharide contents [1]. Out of 8,000 mutagenized clones grown on complete medium (2M+methanol+yeast extract), 11 potential auxotrophic mutants were identified. These isolates were further screened for their specific amino acid requirements, using different combinations of amino acids in a plate. Three mutants required, methionine (met^-), and two leucine (leu^-). Mutants requiring methionine were grown on plate Nos. 4 and 6, and those requiring leucine were grown on plate Nos. 2 and 7 (Tab. 2). The growth requirements for other mutants could not be identified, maybe due to multiple mutations which would require more than one amino acid. However, it is difficult to determine the complete growth requirement with the multiple mutations by the method we used here. Although strong mutagenic conditions were applied, the mutant yield was low. Auxotrophic mutants of obligate methylotrophs have proven difficult to obtain with traditional mutagens [3, 11]. The reason for these difficulties are unknown, but it has been suggested that if the uptake of substances such as amino acids from the medium is complicated, mutations causing auxotrophy would be lethal for the cells unless simultaneous changes occur inside the cells [6].

Transfer of R68.45 to KISR Methylotrophs

The results of the transfer of R68.45 to KISRI methylotrophs are summarized in Tab. 3. The rate of transfer for the tetracycline resistance marker is similar for all four KISRI strains, being approximately 10^{-5} cells/ml. Transconjugants of each cross were checked for the presence of nonselective markers of R68.45 to demonstrate the complete transfer of the plasmid. All transconjugants showed positive growth in the presence of kanamycin (200 μg/ml) and ampicillin (200 μg/ml), whereas the appropriate wild type did not.

Tab. 3. Frequency of transfer of R68.45 from *Ps. aeruginosa* PAO 2001 to KISR methylotrophs

	KISR methylotrophs			
	KISRI-5	KISRI-6.1	KISRI-512	KISRI-5112
Viable count of donor [CFU/ml]	3.7×10^9	2.7×10^9	4.6×10^9	4.9×10^9
Viable count of recipient [CFU/ml]	6.0×10^9	1.4×10^9	2.8×10^9	1.6×10^9
Transfer rate of R68.45 (Transconjugant/recipient cell)	1.5×10^{-5}	3.6×10^{-6}	8.4×10^{-5}	3.5×10^{-5}

The extrachromosomal existence of R68.45 in KISR methylotrophs transconjugants was further demonstrated by agarose gel electrophoresis using R68.45 as positive control and a wild type of KISR methylotrophs as negative control.
Because native plasmids with chromosome mobilizing ability have not yet been found in methylotrophic bacteria [12], we used R68.45 as conjugal plasmid in order to test the recipient ability for the development of the gene transfer methods in KISR methylotrophs. R68.45 was successfully transferred to KISR methylotrophs, although the transfer rate was low. We reproduce the same results several times, with almost the same rate of transfer.

Transfer of R68.45 from KISR Methylotrophs to Ps. aeruginosa

Attempts were made to transfer plasmid R68.45 from KISR methylotrophs transconjugants to *Ps. aeruginosa* ML 4262. Results show that the transfer rate is higher in these crosses than the transfer rate from non-methylotrophs to KISR methylotrophs (Tab. 4). Similar observations were made by ENGEL [6] in other methylotrophs. HOLUBOVA [7] have reported the transfer of incompatibility group P plasmids from *E. coli* and *Ps. aeruginosa* to methylotrophic strains with the low transfer frequency. In all cases the presence of transferred plasmids was confirmed by gel electrophoresis.
Among KISR methylotrophs, KISRI-5 strain was selected for further genetic study, because of the detection of native plasmid and the isolation of specific phage to KISRI-5. The plasmid pSAS was found to be cryptic with 40 kb in size. A specific phage KISRI was isolated from KISRI-5 strain [13]. A detailed study on plasmid pSAS and phage KISRI have been continued in this laboratory.

Tab. 4. Frequency of transfer of R68.45 from KISR Methylotrophs to *Ps. aeruginosa* ML 4262

	KISR methylotrophs			
	KISRI-5	KISRI-6.1	KISRI-512	KISRI-5112
Viable count of Donor (CFU/mL)	4.1×10^9	8.0×10^9	9.9×10^9	1.5×10^9
Viable Count of Recipient (CFU/mL)	4.6×10^9	4.6×10^9	4.6×10^9	4.6×10^9
Transfer Rate of R68.45 (Transconjugant/Recipient Cell)	6.1×10^{-3}	1.8×10^{-2}	1.4×10^{-4}	3.7×10^{-2}

Received June 19, 1989

References

[1] BANAT, I. M., AL-AWADHI, N., HAMDAN, I. Y.: MIRCEN J. App. Microbiol. (1989) (In press).
[2] HAMDAN, I. Y., EL-NAWAWY, A. S., BANAT, I., AL-AWADHI, N.: Process and Microorganisms for producing single cell protein. U.S. Patent Application Serial No. 07/120,322. (Nov. 1987).
[3] O'CONNOR, M. L., HANSON, R. S.: J. Gen. Microbiol. 104 (1978), 105.
[4] TSYGANKOV, Y. D., KAZAKOVA, S. M.: Arch. Microbiol. 141 (1987), 112.
[5] WINDASS, J. D., WORSEY, M. J., PIOLI, E. M., PIOLI, D., BARTH, P. T., ATHERTON, K. T., DART, E. C., BYROM, D., POWELL, K., SENIOR, P. J.: Nature (Lond.) 287 (1980), 396.
[6] ENGEL, J., STEUDEL, A., SATTLER, K.: Acta Biotechnol. 4 (1983), 375.
[7] HOLUBOVA, I., CEJKA, K., KOBEC, N. S., HUBACEK, J.: Folia Microbiol. 32 (1987), 361.
[8] WARNER, P. J., HIGGINS, I. J., DROZD, J. W.: FEMS Microbiol. Lett. 7 (1980), 181.
[9] MILLER, J. H. — In: Experiments in Molecular Genetics p. 431, Cold Spring Harbor, New York, USA. (1977).
[10] DAVIS, R. W., BOTSTEIN, D., ROTH, J. R. — In: Advanced Bacterial Genetics: A Manual for Genetic Engineering p. 209, Cold Spring Harbor, New York, USA. (1980).
[11] MOOR, A. T., NAYDU, M., HOLLOWAY, B. W.: J. Gen. Microbiol. 129 (1983), 785.
[12] HOLLOWAY, B. W. — In: Methylotrophs: Microbiology, Biochemistry and Genetics, p. 87, Ed.: C. T. HOU, Boca Raton: CRC Press Inc. (1984).
[13] AL-MOMIN, S., KADRI, M., MAMAT, U., ENGEL, J. (1989). (Manuscript under preparation).

Acta Biotechnol. 10 (1990) 1, 99–104 Akademie-Verlag Berlin

Heterotrophic Eubacteria Isolated from Cultures of the Cyanobacterium, Spirulina Maxima

DAVIS, J. E., JONES, L. P., ZAJIC †, J. E.

Dept. of Biological Sciences
The University of Texas at El Paso
El Paso, TX 79968

Summary

Three distinct heterotrophic eubacterial strains were isolated from mixotrophic cultures of the filamentous cyanobacterium, *Spirulina maxima* (Gom) Geitl. *Spirulina* spp. are considered to be prime candidates for the phototrophic production of biomass protein, particularly in developing countries. These cyanobacteria are extreme alkaliphiles and halophiles, making their production in arid regions promising.

Most previous studies on the eubacteria which live in *Spirulina* culture systems have focused on determining the possible presence of pathogenic species in biomass protein. Little has been done to understand the symbiotic relationships between the cyanobacterium and its eubacterial cosymbionts. From the perspective of a heterotrophic eubacterium, autotrophic cultural systems of *Spirulina* have limited carbon and energy resources, being limited to cyanobacterial exudates.

In this study, three eubacterial strains were isolated and studied. One strain, a Gram-negative, non-sporing, motile rod, grew exceptionally well in a mineral salts medium where only a small amount of a single low molecular weight organic compound (e.g., acetate) was supplied as sole energy source. This strain was also extremely euryresponsive with respect to salinity and alkalinity as well. Two less well-adapted eubacterial strains are also described.

Introduction

Since the 1964 "discovery" of *Spirulina* [1, 2], several species of the genus *Spirulina* have been studied worldwide for possible utilization as supplemental sources of biomass protein [3]. Although these photoautotrophic organisms are grown at a pH above 8.5 [4–8], and in media containing little added organic material, significant numbers of heterotrophic bacteria are always found in *Spirulina* culture systems [3, 9–12]. These authors have suggested that associated eubacteria use cyanobacterial exudates as their energy sources. Such compounds would likely be low molecular weight substances. OGAWA and TERUI [9] have shown that *Spirulina* does not require the presence of eubacterial cosymbionts in order to grow. In autotrophic outdoor cultivation systems, eubacteria have been reported to occur at 10^5 to 10^6 cells per ml [13]. *Spirulina* can grow mixotrophically with the addition of acetate [14, 15]. The effect of this on eubacterial populations has not been examined, either in terms of diversity or quantity. Previous studies of eubacteria isolated from *Spirulina* cultures have focused on determining whether potential human pathogens are present as part of the microbial load of various cyanobacterial products [9–12].

Little is currently known concerning the physiological and ecological nature of the eubacteria that cohabit *Spirulina* cultures, their relationship to *Spirulina*, or the effects of different *Spirulina* cultural systems on eubacterial populations. This paper attempts to answer some of these questions.

Materials and Methods

Source of Spirulina maxima

A culture of *S. maxima* (Gom) Geitl., UTEX 2342, was obtained from the University of Texas at Austin Culture Collection of Algae. This strain, a Lake Chad isolate from Africa, arrived suspended in a nonaxenic „*Spirulina* medium" [16]. Subcultures were made aseptically into sterile media as described by *Jeeji Bai* and *Subramanian* [17] supplemented with "A6" solution [18] and were incubated at 32 °C in a Conviron Environmental Growth Chamber under a light intensity of 200 W m^{-2} for 12 hours of each 24-hour period. Subculturing was performed at weekly intervals to keep actively growing *Spirulina* available at all times.

Eubacterial Strains

Three strains of eubacteria, designated D-1, D-2 and D-3 were isolated from actively growing mixotrophic cultures of *S. maxima*. Isolation was made on Difco brand Tryptic Soy Agar (TSA), containing 1% NaCl, and the pH adjusted to 9.2. D-1 and D-2 were isolated from a 1% glucose-enriched *Spirulina* medium, while D-3 was isolated from a glucose-acetate-enriched mineral salts medium (0.5%—0.5%). All cultures were maintained on TSA at 32 °C, and were subcultured at appropriate intervals.

Morphology

The morphologies the three isolates were described from colonies growing on TSA and Gram-staining.

Oxygen Requirements

Qualitative evaluation of growth under various gas environments was made according to Slack and Gerenscer [19] using TSA cultures growing at 35 °C.

Preparation of Inocula for Substrate Studies

Isolates of D-1 and D-2 were cultured in 125-ml shake flasks containing 50 ml of Jeeji Bai-Subramanian medium, pH 9.2, supplemented with 1% glucose and 0.5% yeast extract (Gibco). D-3 initially did not grow on this medium, so 0.5% sodium acetate was added. After 24 hours of growth at 32 °C, cells of each isolate were harvested by centrifugation, washed twice in 10 ml of buffered saline (1% NaCl, 0.25% K_2HPO_4, pH 9.2), followed by resuspension in saline.

Nitrogen Requirements

The medium used for preparation of inocula was modified to contain a single source of nitrogen (Tab. 1). The amounts of nitrogen salts used contained equivalent amounts of nitrogen. Growth was monitored using optical density (560 nm) measurements on a Bausch and Lomb Spectronic 20 for cultures at 32 °C. The initial pH was 9.2—9.4. Cultures were shaken by hand four times a day.

Tab. 1. Substrates for nitrogen utilization by eubacterial isolates

Nitrogen source	[gl^{-1}]
Ammonium chloride	1.6
Ammonium nitrate	1.2
Vitamin-free casamino acids (Difco)	1.0
Sodium nitrate	2.5
Urea	0.9
Yeast extract, Gibco	1.0

pH Tolerance

Growth of each isolate was monitored in a series of tubes containing Difco brand Tryptic Soy Broth (TSB), the pH's of which were adjusted to 3.8, 4.7, 5.4, 6.4, 7.2, 8.1, 8.6, 9.6 and 10.6, respectively, using sterile 1N HCl or 2N NaOH. Incubation was at 32 °C with agitation provided by handshaking four times a day.

Sodium Chloride Tolerance

Sodium chloride tolerance was evaluated in shake flask cultures of the basal medium (See preparation of inocula), pH 9.2. Levels of NaCl used were 1, 5, 9, 13, and 17%, respectively. Incubation was at 32 °C.

Temperature Tolerance

Growth of each isolate was evaluated by cultivation in TSB, pH 9.2, at 5, 15, 25, 35, 45, 55 and 65 °C. Cultures were agitated four times daily.

Fermentation of Carbohydrates

Difco brand Phenol Red Carbohydrate Fermentation Media was used to test each isolate's ability to produce acid and/or gas from the following substrates: sodium acetate, arabinose, fructose, galactase, glucose, lactose, maltose, mannitol, rhamnose, sucrose and xylose. Incubation temperature was 35 °C, and pH after sterilization was adjusted to 9.2.

Miscellaneous Tests

Presence of catalase was determined by dropping 3% H_2O onto colonies growing on TSA. Urease production was determined on Urea Medium (Difco) for cultures growing at 35 °C. Oxidase tests also performed.

Use of Acetate, Butyrate and Palmitate

Each isolate's ability to utilize fatty acids as sole energy sources was tested in the basal medium (pH/9.2) at 35 °C in shake cultures.

Sensitivity of D-1 to Different Amounts of Acetate

The effects of varying acetate concentration on the D-1 strain was studied because it was the dominant strain occurring in acetate-enriched cultures of *S. maxima*. Shake flask cultures of basal medium contained 0.01, 0.02, 0.1, 0.2, 1.0 and 2.0% acetate as the only energy source were grown at 35 °C. In addition to the original amount of acetate present, an equal amount of acetate was added after 24 hours of growth. Dilution plating was used to estimate eubacterial numbers in relation to acetate feeding.

Growth on Other Organic Molecules by D-1

Shake flask cultures containing basal medium, and 0.1% of the following compounds as the only energy source inoculated with D-1, and incubated at 35 °C: methanol, ethanol, n-propanol, isopropanol, formaldehyde, acetaldehyde, propanaldehyde, formic acid, sodium propionate, vitamin-free casamino acids (Difco), β-aminobutyrate, 3-hydroxybutyrate, 4-hydroxybutyrate, isobutyrate, isovaleraldehyde, β-ketobutyrate, sodium lactate, oxalacetate, tryptophan, alanine, aspartate, cystine, glutamic acid, glycine, homoserine, isoleucine, leucine threonine, and valine.

Results

The general characteristics of three eubacterial isolates are given in Tab. 2. None of the three isolates was a sporeformer. Isolate D-1 grew well on JEEJI BAI-SUBRAMANIAN Spirulina medium supplied with glucose and inorganic nitrogen, but this isolate also grew well on basal medium containing each of the following compounds added singly as the sole energy source: ethanol, n-propanol, propionate, vitamine-free casamino acids, 3-hydroxybutyrate (the substrate for polyhydroxybutyrate, *Spirulina's* chief storage product), isobutyrate, isovaleraldehyde, lactate, oxalacetate, aspartate, alanine, glutamate, butyrate, palmitate and acetate. Non-utilized substrates included methanol, isopropanol, acetaldehyde, formaldehyde, formate, β-aminobutyrate, 4-hydroxybutyrate, β-ketobutyrate, tryptophan, cysteine, glycine, homoserine, isoleucine, leucine, threonine, and valine. Any nitrogen sources could be utilized.

Tab. 2. Characteristics of three eubacterial isolates from *Spirulina maxima* cultures

	D-1	D-2	D-3
Cell morphology	bacillus 2.0 × 0.6 μm singly & clumps	bacillus 1.0 × 0.7 μm filaments, pairs, & v-shapes	coccus 1.0 × 1.0 μm pairs, tetrads, & clumps
Colonies on TSA	creamy white smooth, entire	red smooth, flat	yellow to orange smooth, flat
Gram reaction	negative	positive	positive
Motility	yes	no	no
Catalase	positive	positive	positive
Oxidase	positive	positive	positive
Urease	positive	negative	positive
Gas requirement	aerobic anaerobic	aerobic anaerobic	aerobic
pH range	7.0–10.6	7.0–9.5	5.4–9.5
NaCl tolerance	0–13%	0–5%	0–5%
Temperature	15–45 °C	15–35 °C	15–45 °C
Acetate	yes	no	yes
Butyrate	yes	no	yes
Palmitate	yes	no	no
Acid from:	arabinose fructose galactose glucose maltose mannose sucrose	arabinose fructose galactose glucose lactose maltose mannitol sucrose	arabinose fructose galactose glucose maltose mannose sucrose

Strains D-2 and D-3 utilized all nitrogen sources (except urea for D-2), however, best growth occurred in the presence of organic nitrogen.
In a separate unpublished study comparing the effects of acetate supplementation on *Spirulina maxima* growth, total eubacterial numbers was consistently 10^7 ml^{-1} in the absence of of acetate with the D-1 strain predominant. When small amounts of acetate were added, eubacterial number increased to 10^{10} ml^{-1}, and only the D-1 isolate was recoverable.

Discussion

Until the present study, the biological and ecology properties of the eubacteria living as cosymbionts in cultures of the alkaliphilic cyanobacteria *Spirulina* have not been closely examined. Most frequently isolated species from cultural systems have included *Bacillus*, *Micrococcus*, enterococci, as well as Gram-negative rods [9—12].
Strain D-1 did share some physiological characteristics with certain strains of WATANABE et al. [20], and HORIKOSHI and AKIBA [21], but could not be considered a pseudomonad, as D-1 is a facultative anaerobe. Nor does it closely resemble any genus already described within the Enterobacteriaceae [22]. This organism was extremely euryresponsive [21] in terms of many environmental parameters, and utilization of small molecular substrates for energy.
Strain D-2 differed from the coryneform alkaliphilic isolates of HORIKOSHI and AKIBA [21], and KOBAYASHI and HORIKOSHI [23] in that this strain possessed red colonies, could not grow above a NaCl concentration of 10%, had a narrow range of pH tolerance, could use ammonium salts as sole nitrogen source, and grew anaerobically. Carbohydrate fermentation patterns also differed considerably. GEE et al. [24] reported on five strains of alkaliphilic corynebacteria. However, their description made no mention of v-shaped morphologies. In addition, their isolated could not ferment arabinose or lactose, nor could they use ammonium or nitrate. SOUZA and DEAL [25] have described a new alkaliphilic coryneform strain, but their isolate is a strict aerobe, without v-shaped morphology.
Strain D-3 closely resembled the *Micrococcus* strain 31-2 of HORIKOSHI and AKIBA [21], as well as the description given by BAIRD-PARKER [26].
In unpublished experiments comparing autotrophic and mixotrophic (acetate) growth of *Spirulina maxima*, the total number of eubacteria in autotrophic cultures was very similar to previously published reports (10^7 ml^{-1}), however, strain D-1 was much more numerous than either D-2 or D-3. When acetate was added, eubacterial numbers increased three-fold (10^{10} ml^{-1}), and only the D-1 isolate was recovered.
In conclusion, D-1 and D-2 may represent new taxa. The taxonomic position of many alkaline isolates is difficult to ascertain because their physiology is probably quite different from their relatives in more moderate environments. As alkaline environments are more extensively studies, it can be expected that many new taxa will be discovered.

Received June 19, 1989

References

[1] ZARROUK, C.: Ph. D. Thesis, Universite' de Paris, 1966.
[2] LÉONARD, J., COMPÉRE, P.: Bull. Jard. Bot. Nat. Belg. **37** (1967), 1.
[3] CIFFERI, O.: Microbiol. Rev. **47** (1983), 551.
[4] JOHNSTON, H.: Tutara **18** (1970), 19.

[5] MELACK, J.: Limnol. Oceanogr. 24 (1979), 753.
[6] MELACK, J., KILHAM, P.: Limnol. Oceanogr. 19 (1974), 743.
[7] PANKOV, K., ERGASHEV, A., SAADI, A.: Uzb. Biol. Zh. 0 (1980), 39.
[8] TUITE, C.: FRESHW. Biol. 11 (1981), 345.
[9] OGAWA, T., TERUI, G.: J. Fermnet. Technol. 48 (1970), 361.
[10] JACQUET, J.: Ann. Nutr. Alim. 29 (1976), 589.
[11] BECKER, E., VENKATARAMAN, L.: Biotechnology and Exploitation of Algae: the Indian Approach. Germany Agency of Technical Cooperatin, Eschborn, F.R.G., 1982.
[12] BECKER, E.: Microalgae, Findings of Three Experimentation Projects. Germany Agency for Technical Cooperation, Eschborn, F.R.G., 1983.
[13] RICHMOND, A., KARG, S., BOUSSIBA, S.: Plant Cell Physiol. 23 (1982), 1411.
[14] BARON, C., FORIN, M.: C. R. Seances Soc. Biol. Dij. 171 (1976), 73.
[15] CRANCE, J., FORIN, M., BARON, C.: C. R. Acad. Sci. Paris 284 (1977), 89.
[16] SCHLOESSER, U.: Ber. Deutsch Bot. Ges. 95 (1982), 186.
[17] JEEJI BAI, N., SUBRAMANIAN, S. — In: Proc. Nat'l. Workshop Algal Systems, 1980.
[18] ARNON, D.: Amer. J. Bot. 25 (1938), 322.
[19] SLACK, J., GERENSCER, M.: Actinomyces, filamentous bacteria: biology of pathogenicity. Burgess Publishing Co., Minneapolis, 1975.
[20] WATANABE, N., OTA, Y., MINODA, Y., YAMADA, K.: Agric. Biol. Chem. 41 (1977), 1353.
[21] HORIKOSHI, K., AKIBA, T.: Alkalophilic microorganisms, a new microbial world. New York: Springer-Verlag, 1982.
[22] COWAN, S. T. — In: Bergey's Manual of Determinative Bacteriology. 8th Ed., R. BUCHANAN and N. GIBBONS (eds.). Baltimore, Williams and Wilkins Co., 1974, pp. 290.
[23] KOBAYASHI, Y., HORIKOSHI, K.: Agric. Biol. Chem. 44 (1980), 41.
[24] GEE, J., LUND, B., METCALF, G., PEEL, J.: J. Gen. Microbiol. 117 (1980), 9.
[25] SOUZA, K., DEAL, P.: J. Gen. Microbiol. 101 (1977), 103.
[26] BAIRD-PARKER, A. C. — In: Bergey's Manual of Determinative Bacteriology, 8th Ed., R. BUCHANAN and N. GIBBONS (eds.). Baltimore, Williams and Wilkins Co., 1974, pp.

Book Review

M. S. S. PAIS, F. MAVITUNA, J. M. NOVAIS

Plant Cell Biotechnology
(NATO ASI Series H: Cell Biology, Vol. 18)

Berlin, Heidelberg, New York, London, Paris, Tokyo: Springer-Verlag, 1988
500 pp., 92 figs., DM 298.00, ISBN 3-540-18556-9

Plant Cell Biotechnology offers significant potential benefits in the areas of plant improvement and production of fine chemicals. The book is the Proceedings of a NATO Advanced Study Institute held in Portugal 1987 and includes the review and research lectures given, papers based on poster presentation of current research and notes on some of the discussion sessions. It covers many of the diverse topics in this field: applications in agriculture and plant breeding, plant genetic engineering, biochemical and engineering aspects of biotransformation and production of the fine chemical by plant cell and tissue cultures. A big part of the book contains the large-scale cultivation of plant cells, bioreactors of plant cell culture and engineering aspects. That's why the book should appeal to the researchers, both experts and novice in the field, and professionals from agricultural, pharmaceutical and chemical sectors with a current or potential interest in what plant cell biotechnology can offer.

R. PÄTZ

Acta Biotechnologica

Volume 10 1990 Number 1

Contents

Acta Biotechnologica is indexed or abstracted in Current Contents/ET & AT; Chemical Abstracts; Biological Abstracts; Biotechnology Abstracts; Excerpta Medica

Bei Fragen zur Produktsicherheit wenden Sie sich bitte an:
If you have any questions regarding product safety,
please contact:

Walter de Gruyter GmbH
Genthiner Straße 13
10785 Berlin
productsafety@degruyterbrill.com